GROUNDWATER AND SOIL REMEDIATION: PRACTICAL METHODS AND STRATEGIES

by

Evan K. Nyer

ARCADIS GERAGHTY & MILLER

Ann Arbor Press
Chelsea, Michigan

Library of Congress Cataloging-in-Publication Data

Nyer, Evan K.
    Groundwater and soil remediation : practical methods and strategies /
by Evan Nyer.
        p.   cm.
    Includes bibliographical references and index.
    ISBN 1-57504-088-3
    1. Groundwater--Purification. 2. Soil remediation. I. Title.
TD426.N9397   1998
628.1'68--dc21   98-2874
                                                                        CIP

ISBN 1-57504-088-3

ANN ARBOR PRESS
121 South Main Street, Chelsea, Michigan 48118
Ann Arbor Press is an imprint of Sleeping Bear Press

PRINTED IN THE UNITED STATES OF AMERICA
10 9 8 7 6 5 4 3 2 1

INTRODUCTION

This book is a compilation of my columns that have been published in *Ground Water Monitoring and Remediation* over the last five years. This is the second time that I have put five years' worth of articles together. As before, I felt that it was important to put all of these papers together so that readers would have a convenient place to refer to the entire set.

The papers cover many subjects from commentary to advanced technologies. The one theme that holds true for all of the them is cost savings. I have used costs as a basis to compare between technologies and to present practical approaches for the entire remediation process. The papers are organized into five sections: The Basics—The Only Way to Save Real Money; Pump and Treat Remediation; Working with Regulators; Biological Remediation; and Advanced Remediation Techniques. The papers within each section are arranged chronologically.

The reader will find these papers a little different than the normal technical paper in the field of remediation. While the objective is to transfer knowledge in order to improve remediation designs, the methodology is Lite. No, not intoxicating (you have to see me in person for that), but not highly technical. This has allowed me to concentrate on concepts instead of rigorous data presentation. It has also allowed me to summarize other people's work. A great example of this is Chapter 17, "The State-of-the-Art of Bioremediation," in which I was able to pull together several important papers on bioremediation in one place. Since I have not been thrown out yet, the readers must find this style and information helpful. Hopefully, you will all find this book easy to read and helpful also.

It is hard to believe that I have been writing "Treatment Technology" for over 10 years. The one constant during all of that time is that the only way I could have continued producing quality articles was with significant help. That help has taken many forms. First, the National Ground Water Association has been very supportive. Anita Stanley, who has been editor of *Ground Water Monitoring and Remediation* for my entire tenure, and the rest of the organization have allowed me to pick my own subjects and maintain a lighter style of writing. This confidence and support has produced the articles that you will find in this book.

Next, I have worked for Geraghty & Miller for most of the time that I have been writing this column. All of the papers in this book have had great support from Geraghty & Miller, Inc. This support has come in the form of graphical and secretarial services, technical resources and reviews, and time in order to write the papers.

Finally, but most importantly, I have had help from all of my co-authors. You cannot keep a full-time job and write a regular column without help. They have also been responsible for the continuous high quality of the articles. Each chapter lists the co-authors in the title section. The back of the book gives a brief biography on each co-author. Other people have had input, and I have tried to mention them in the article itself. This includes the references listed in each article. I always felt that it was my place to promote major concepts. Sometimes, I have used other published work for the details to fill out the concepts. While I always tried to give these authors full credit, I also want to thank them for their help in my work.

Let us begin this book with one of my favorite, fun articles.

AUTHOR BIOGRAPHIES

- **Evan K. Nyer** is a Vice President with Geraghty & Miller, Inc., where he is responsible for maintaining and expanding the company's technical expertise in geology/hydrogeology, engineering, modeling, risk assessment and bioremediation. He has been active in the development of new treatment technologies for many years. He has been responsible for the strategies, technical design, and installation of more than 300 groundwater and soil remediation systems at contaminated sites throughout the United States.

  Mr. Nyer received his graduate degree in environmental engineering from Purdue University and has authored three books, *Practical Techniques for Groundwater and Soil Remediation,* published by Lewis Publishers, *Groundwater Treatment Technology,* now in its second edition, published by Van Nostrand Reinhold, and *In Situ Treatment Technology.* Mr. Nyer is a regular contributor to *Groundwater Monitoring and Remediation* having had his own column, "Treatment Technology," in the periodical for the past 10 years.

- **Gary Boettcher** has 10 years of environmental experience obtained in the chemical industry, decontamination equipment manufacturing, hazardous waste treatment industry, and environmental consulting. He received a Bachelor of Science in Microbiology in 1984 from the University of South Florida and completed graduate course work in public health. Mr. Boettcher specializes in investigation and remediation of impacted groundwater and soil. His technical expertise focuses on evaluation of physical, chemical, and biological remedial alternatives. Mr. Boettcher has been a project engineer, scientist, and manager on federal and state Superfund, RCRA, and various industrial projects throughout the United States, including the Bahamas and Puerto Rico. Mr. Boettcher has developed added specialization in the area of bioremediation where he has designed, managed, and implemented in-situ and ex-situ bioremediation processes to treat hydrocarbons and industrial solvents. Mr. Boettcher has co-authored several papers on remediation and contributed to the development of U.S. Environmental Protection Agency (USEPA) guidance documents focusing on use of aerobic biological treatability studies at CERCLA sites.

- **Michael E. Duffin** is an environmental hydrogeologist/project manager with the firm of Geraghty & Miller, Inc., Austin, TX. He graduated with B.S. and M.S. degrees in geology from Baylor University, and received his Ph.D. in geochemistry from the University of Illinois. Duffin is a registered professional geologist, and has worked for the last eight years in the environmental consulting field managing waste site investigation and remediation projects.

- **Edward G. Gatliff** is President of Applied Natural Sciences, Inc. in Hamilton, Ohio. He was formerly research director for agricultural studies at Servi-Tech, Inc., and has directed the analytical sciences division for Bowser-Morner, Inc. In addition to the development of the TreeMediation process, he has worked extensively with remote monitoring using computers and sensors. He continues to conduct research and consultation services for TreeMediation and other vegetative remediation projects.

- **Mary Gearhart** is a Principal Engineer in Geraghty & Miller's Seattle office and has more than 17 years of experience in environmental engineering. She has managed or been involved in nearly 100 groundwater investigations and remedial actions, including those with solvents, wood product wastes, low-level radioactive wastes, and inorganics from industrial processes. She is currently serving as senior advisor for several litigation support efforts regarding groundwater plumes and subsequent remediation plans.

- **Pete Jalajas** is a Senior Scientist at Geraghty & Miller, Inc. He specializes in remedial engineering project management, litigation support, geochemical dating and fingerprinting, and the development of innovative strategies and technologies for environmental assessment and monitoring. He has a Bachelor's degree in Geochemistry from Occidental College and a Master's degree in Geochemistry from the University of Southern California; he is also an Engineer-in-Training. His e-mail address is pjalajas@gmgw.com. Address: Geraghty & Miller, Inc., One Corporate Drive, Andover, Massachusetts 01810. Phone: (508) 794-9470. Fax: (508) 682-4452. Homepage: http://www.gmgw.com.

- **Bridget Morello** received a B.S. in Chemical Engineering from the University of South Florida in 1987 and is a Project Engineer, EIT with Geraghty & Miller, Inc. in Tampa, Florida. She is responsible for managing several multimillion dollar projects and has extensive experience in the design of groundwater and soil treatment systems and waste minimization processes, including bid and contract documents, construction oversight, and O&M.

- **Deepak Nautiyal** is a Staff Engineer with over six years of experience in the applied Civil Engineering field, with three years specifically in the site remediation application of Civil/Environmental Engineering. He specializes in remedial design and pneumatic fracturing, and his primary responsibilities are providing engineering support to environmental projects. Mr. Nautiyal has a B.S. in Civil Engineering from the Institute of Technology, India, an M.S. in Building Engineering & Management from the School of Planning & Architecture, India, and an M.S. in Environmental Engineering from the New Jersey Institute of Technology.

- **Terry Regan** is an Associate and Principal Scientist in Geraghty & Miller's Andover, Massachusetts office, where he is currently the office manager. His primary technical responsibilities have included managing large-scale environmental investigation and remediation projects. Mr. Regan has a B.S. in Environmental Science/Geology from the University of Lowell, and an M.S. in Hydrology from the University of New Hampshire. He has over 13 years of experience in the environmental industry.

- **Gregory J. Rorech** is a Technical Director of Engineering with Geraghty & Miller, Inc., where he specializes in the evaluation, design, development, and implementation of both conventional and innovative remediation technologies. Mr. Rorech is responsible for training, development, and implementation of remedial technologies within Geraghty & Miller. Remedial technologies recently implemented include in-situ biological remediation, phytoremediation, air sparging, biosparging, reactive walls, enhanced vapor extraction, vacuum extraction, reverse osmosis, ion exchange, land farming, air stripping, liquid and vapor phase carbon absorption, thermal and catalytic oxidation, trickling filter, advanced oxidation processes, electrochemical precipitation, and iron removal. He is a contributing author to four books and has written extensively on groundwater and soil remediation technologies. He is a frequent lecturer on environmental remediation at the Princeton Remediation Course, Groundwater Contamination & Remediation Techniques seminar, and Florida Chamber of Commerce Environmental Permitting Short Courses.

- **David C. Schafer** is a principal scientist/associate with Geraghty & Miller, Inc. He attended the University of Minnesota where he obtained a Bachelor's degree in mathematics and a Master's degree in computer science. Prior to joining Geraghty & Miller, he worked for a major water well equipment manufacturer for 19 years. Throughout his career, he has specialized in the design of monitoring and production wells and the analysis and interpretation of pumping test data. He has written numerous articles on these subjects and has designed and analyzed pumping tests from hundreds of wells throughout North America.

- **Charles D. Senz**, a senior geologist with Geraghty & Miller, Inc. in Denver, Colorado, has more than 14 years of experience as a geologist including 8 years in the environmental field. He has managed projects, performed extensive field work, and written numerous reports for various hazardous waste sites, CERCLA and RCRA sites, and hydrocarbon-contaminated sites in the Rocky Mountain region. Mr. Senz has experience at sites impacted by volatile and semivolatile organic compounds, chlorinated solvents, petroleum hydrocarbons, PCBs, pesticides and herbicides, metals and other inorganic compounds, and acids and strong bases. He has also conducted and managed environmental site assessments for residential properties, small and large industrial properties, warehouses, hotels, office buildings, natural-gas production facilities, medical facilities, photo-processing facilities, and large tracts of vacant land.

- **Lynne Stauss** is an Environmental Engineer at NFESC (Naval Facilities Engineering Service Center), has a B.S. in Geology, Edinboro University, Edinboro, PA, an M.S. in Geology, Western Washington University, Bellingham, Washington, 5 years as petroleum geophysicist (UNOCAL), and 5 years R&D innovative remedial technologies for hazardous waste sites (Naval Civil Engineering Laboratory).

- **Suthan S. Suthersan** is a Vice President and Director of Remediation Technologies at Geraghty & Miller, Inc. His primary responsibilities include development and application of innovative in-situ remediation technologies and providing technical oversight on projects across the entire country. He has a Ph.D. in Environmental Engineering from the University of Toronto and is also a registered Professional Engineer in several states. His technology development efforts have been rewarded with many patents awarded and pending. Dr. Suthersan has pioneered various conventional and modified applications of many in-situ technologies such as in-situ reactive walls, in-situ air sparging, bioventing, in-situ bioremediation and pneumatic/hydraulic fracturing. His primary strength lies in developing the most cost-effective site-specific solutions utilizing the latest cutting edge techniques. He has developed a national reputation in convincing the regulatory community to accept the most innovative remedial techniques.

CONTENTS

Introduction

Part I
The Basics—The Only Way to Save Real Money

Part II
Pump and Treat Remediation

Part III
Working With Regulators

Part IV
Biological Remediation

Part V
Advanced Remediation Techniques

INTRODUCTION

1

A WISH LIST

Evan K. Nyer

THE HOLIDAYS ARE ONCE AGAIN quickly approaching, and I know all of you have been wondering what to get me for Christmas. Instead of going through the embarrassment of returning unwanted gifts this year, I have decided to produce a wish list of things that I would like for Christmas. You will notice that none of the items have any monetary value, so there should be no tax consequences from any of these gifts.

The following are the things that I would like:

A New Method for Reporting Analytical Data

I PROPOSE THAT WE COME up with a new method for presenting analytical data in reports and tables. While we all know that to really understand what a number means when we present it in association with the concentration of a contaminant, we have to understand the QA/QC that goes along with that number. The problem is that most of the time, the only thing that is put into the report is the specific number. Therefore, I would like to change the method in which we put the numbers into the report. When we are very sure of the numeric value of the concentration, the numbers should be reported in black. As we lose confidence in the numeric value, then we should lighten the printing of the number. For example, most values of parts per million would be printed in black; most values of parts per billion would be printed in gray. Values reported in parts per quadrillion would be reported in white. This system could also be used when we are not sure of the sampling method or there are a lot of organics that could interfere with the analysis, even at high concentrations. The method would be simple and the reader could immediately see what type of confidence there was in the number that is presented in the report. I also think that the public could quickly understand this method, even though the white printing on a white piece of paper may confuse some.

Outlaw the Term "Clean"

THERE IS NO SUCH THING as clean (I think that I remember having a similar discussion with my mother when I was much younger). Clean is a relative term. We can get conditions at a site below certain concentrations, but we are never really able to completely eliminate all of the molecules of a contaminant once it has been at a site. As methods to detect lower concentrations continue to be developed, we will begin to find these compounds again. Therefore, we can never say that we have eliminated this compound from a site. The problem with the term "clean" is it gives the public the notion that the compounds have been eliminated. Even the MCLs are based upon the limitation of cancer risks, not the elimination of cancer risks.

Therefore, I would like to switch to a scale based on risk to describe sites. We could then compare this scale to things like smoking cigarettes, sunbathing, and other everyday occurrences so that the public could understand what we were trying to accomplish. Removing the barrels from a site would change the risk from smoking two packs of cigarettes per day to smoking one cigarette per day. The addition of a fence around the site would lower the risk to sunbathing without suntan lotion. As more compounds were removed or controlled, the related risk would decrease. Once all of the steps of the remediation were complete, the risk could be lowered to say, working at a post office. With risk assessment being a hot topic in Congress this year, I might actually have a chance of receiving this wish.

Get the EPA Out of the Design Business

I BELIEVE THAT THE EPA and the other regulatory agencies have a very important part in the cleanup of contaminated sites. However, I cannot express to you my feelings when a recent graduate of an engineering school tells me that they disagree with my design and I have to change it. I am very proud of myself when I am able to continue talking in a normal voice when this happens. I understand their detailed analysis of analytical data and sampling methods to ensure that the data collected represents the site. But I do not understand the regulatory agencies making decisions on specific designs of treatment methods. The regulators should concentrate on making sure responsible parties respond to contaminated areas in a timely manner. I know that this is a tough wish, and that the EPA will tell me that they need to be in the design business because of all the bad designs that are being implemented. However, I thought that that was my job.

Two New Research Areas

THERE ARE TWO AREAS OF research that I would like to see emphasized. First, I would like to see more work on the microenvironment as opposed to the macroenvironment. We have tremendous amounts of knowledge on the effect of clay and sand lenses and other geologic formations on the movement of plumes. However, we need to gather a lot more information on the effect of the microenvironment on the remediation of a site. My geologist friends have tried to explain 'geologic' time and how it makes geologists regard the world a little differently than the rest of the human race. However, I am not aware of any thing like 'geologic' size that would prevent the evaluation of the microenvironment.

Geologists have been well trained in plume movement, and the effect of macrogeological conditions on that movement. But, wake up, this is the '90s. We are remediating sites now and we have to shrink the plume. We need to develop a better understanding of how the microenvironment controls the remediation through diffusion, and then teach this to our geologists who are in the business of remediating sites. As I have stated many times before, geologists should be in charge of in-situ remediations. But they will have to have the knowledge available before they can run these projects effectively.

The next area of research is in the Biochemical area. We seem to have a tendency to want to study specific bacteria. For some reason, being able to name the bacteria that might be degrading the compound in the field seems important to people. I think that this is left over from the belief that biological remediations are somehow connected to magic. I would like us to switch to understanding the environment in which bacteria flourish and degrade compounds. I think that the environmental studies would get us a lot further in our understanding and acceptance of the biological technologies. This work has already started with the increased use of intrinsic biological remediation at many sites. It is a pleasure to see the study of intrinsic reactions is based upon the biogeochemical environment and not a specific bacteria.

Someone to Explain to Me Why Geologists Always Need 'Just A Few More' Wells in Order to Understand a Site

I HAVE NEVER BEEN IN a situation where the new geologist at a site was satisfied with the old well pattern. I am the first to say that the key to a good remediation design is a complete understanding of the geology of the site and the plume delineation. But is there a class at Geology School that teaches that you are not really a geologist unless you are drilling? Or is it a territory thing? It is not your site until you mark it with your well? Maybe I am bringing this up because I miss the days when geologists and engineers used to argue. Things have been too quiet lately. People will think that we are getting along.

An Engineer That Would Provide Accurate Costs

(I MAY NOT BE NICE, but I am fair). With 10 to 20 thousand groundwater treatment systems installed across the country, you would think that the data would be available from which an engineer could provide the actual costs of a remediation. It is hard to believe that it is still necessary to cost equipment ± 50%. This is not a small matter. We plan to spend $3–5 billion on remediation during the next few years. If that figure is ± 50%, then either we do not have enough or we have too many remediation engineers. We don't accept bids from contractors or equipment vendors at a range of costs. Why do we accept this from engineers? I think that engineers are basically afraid of being wrong, and these ranges are just a method to ensure that the engineer can say that he/she was right at the end of the project. Try stopping a car ± 50 feet of the stop sign.

Lawyers

I THINK THERE SHOULD BE a method in which the lawyers get some direct response from working on a remediation project. One of the ways in which we could do this is to simply have the lawyers feel mental anguish that grows in intensity the longer they spend on a particular project. The lawyers themselves have come up with a similar method in that the responsible party feels pain based on the cost from the lawyer being on the project. The longer the lawyer works on the project, the more the monetary pain that the responsible party must feel. This method would simply make the process more direct. The old process has not seemed to work anyway; 30% of the money spent on Superfund went to lawyers. Of course, any actual improvement in the site condition that the lawyers facilitated with their time would reduce the anguish that they would experience. This will also relieve most engineers and geologists from any connection to this mental anguish formula.

A New Law That Limits the Size of Reports

I THINK THAT ONE OF the best ways to save money on environmental remediations would be to limit the size of the reports that go along with the projects. Sometimes it seems that we have developed an entire corps of geologists and engineers that have only been trained in report writing. Actual geology, hydrogeology, and engineering make up a small portion of the time spent by our young professionals. If you add in review time by local, state, and federal regulator, the total time spent on just paper is amazing.

The problem is that I am not sure that all of this paper makes the remediation any better in the end. We end up thinking that 'big' projects need big reports or the clients will not think that they have gotten their money's worth. We need to break the cycle. The only way is for the federal government to produce an amendment to Superfund legislation. I know that this is a radical step, but I do not think that the various parties have the strength of will without the political cover of an amendment.

I realize that we could not simply limit the size of all reports. Various projects require different amounts of information. I propose that the amendment be based upon the use of block grants. Each state would receive a total amount of pages per year and it would be up to the state to allocate these pages to each project.

This is a very simple proposal, and if it is presented in the correct manner to Congress, it has a good chance of passing. We need this legislation to save time, money, and the careers of our young professionals. (By the way, I am very proud that I did not have to resort to the 'saving trees' argument to get my point across.)

Writer's Compensation

AUTHORS NEED TO BE MORE fairly compensated for the written word. It seems that we are willing to spend a great deal of money to hear someone talk, but the written word should be free. I propose that all of you send me a small compensation per word of my articles that you read (each picture is 1,000 words). We can work on the tax consequences later.

As with all Christmas lists, I do not expect to get everything (especially since I am Jewish). But this article has come out early enough for everyone to put forward a good effort. We'll go back to a technical article next time. I am allowed to have fun every once and a while.

PART I

THE BASICS–THE ONLY WAY TO SAVE REAL MONEY

2

ALL SITES CANNOT BE "CLEAN"

Evan K. Nyer and Douglas Hatler

THE MOST IMPORTANT PHASE OF a remediation project is developing the strategy. A good strategy will save more money than implementing the most advanced treatment technology on a project. The good news is that in most circumstances we have sufficient experience to review preliminary site data and develop a strategy. The bad news is that we often write the Scope of Work before we discuss the strategy. Too often we start the site investigation prematurely. After COMPLETELY delineating the contamination plume and defining the aquifer characteristics, we begin to feel comfortable with our understanding of that specific site. A few hundred thousand dollars later, we begin to evaluate the proper treatment technologies. After completing this task, we select the best treatment method, build the system, and proceed to pump and treat.

This order of events is expensive and procedurally incorrect. By treating the site remediation like a "cookbook," we have ignored all the knowledge that we gained over the past 10 years from other site remediations. While each site is unique, there are many techniques that we can transfer to other sites.

We currently have access to information derived from groundwater treatment systems that have been running from five to ten years. These long-term treatment systems provide data on the actual effect of pumping on various contaminant types and aquifer systems. We have the opportunity to change our investigation and remediation methods so that this knowledge is incorporated into the process.

The first mistake that is typically made is selecting the wrong project objective. Too often, we simply set the objective "to clean the site." The problem with setting this objective is that one of the main lessons we have learned over the past five years is that we cannot reach "clean" at all sites. Sites with nondegradable organic contamination do not reach "clean." Our best method for addressing contaminated zones of an aquifer, at the present time, is pump and treat. While this method is a very powerful tool to control plume movement, and to ensure that there is no human contact with contaminated groundwater, there is broad evidence that pump and treat is not a successful cleaning method for aquifers.

This is not just the opinion of the authors. Since 1989, several people have been publishing information on the limitation of pump and treat. We will first review some of the major articles that have been written on this subject. Then, to further support this point, we will review a specific site and put a dollar figure to setting the wrong objective.

Previous Work

SINCE 1989 THERE HAVE BEEN several detailed studies on the efficacy of pump and treat to remediate Superfund sites. The original work was done by the EPA and published in 1989 (USEPA, 1989). The paper was entitled "Evaluation of Groundwater Extraction Remedies." This study evaluated 19 sites and reported that 13 sites had aquifer restoration as their primary goal. The conclusion was that only one of the sites was successful in remediation so far. These numbers provided great headlines for several magazine articles. While these statistics made the report exciting, the detailed conclusions are very important. The paper outlines three main conclusions:

1. The groundwater extraction systems were generally effective in maintaining hydraulic containment of contaminant plumes, thus preventing further migration of contaminants;
2. Significant removal of contaminant mass from the subsurface is often achieved by groundwater extraction systems. When site conditions are favorable and the extraction system is properly designed and operated, it may be possible to remediate the aquifer to health-based levels; and
3. Contaminant concentrations usually decrease most rapidly soon after the initiation of extraction. After this initial reduction, the concentrations often tend to level off, and progress toward complete aquifer restoration is usually slower than expected.

This study was followed by several other reviews. While the statistics were useful, it is important to get to the reasons behind the statistics if we were going to be able to successfully design pump and treat systems. In October of 1989, a technical review by Joseph Keely, titled "Performance Evaluations of

Pump and Treat Remediations" (J. Keely, 1989), was published. This work went into the technical details on the limitations of pump and treat systems. It showed why we have a significant removal of mass in the beginning of a pump and treat. It also showed why the mass removal declines rapidly and then approaches a steady state level. Several times this column has referred to Keely's paper as a must-read for anyone in the groundwater restoration industry.

Several authors outside of the EPA have picked up this theme during the last several years. Every major magazine in the environmental area has had at least one article on the limitations of pump and treat. The titles always seem to concentrate on the failure of pump and treat to be able to completely remediate a site. The technical portion of the paper, however, usually provides good information and insight into the proper application of pump and treat, as a part of a remediation.

The EPA is now taking all this information to heart, and EPA Assistant Administrator Donald Clay has been quoted, "where it is determined that it is not practical to restore portions of contaminated ground water with currently established technology, due to the presence of DNAPL's, alternative actions, including containment and shrinking of the contaminated area should be implemented." (D. Clay, 1991). From this quotation it can be concluded that the EPA now realizes that pump and treat is an important prescription for remediation, but not the exclusive cure.

The federal EPA has gone further and has recently announced in the February 14, 1992 *Federal Register* (*Federal Register,* 1992), a request for quotations for technical organizations to submit proposals on research leading to practical methods for enhancing the effectiveness of pump and treat systems. They stated that the processes should have an emphasis on improving contaminants extraction from the aquifer.

Once again, private industry is paralleling EPA's work in this area. Our own National Ground Water Association is planning an educational program this September 30 through October 2, 1992, titled "Aquifer Restoration: Pump and Treat and the Alternatives." This program will be in conjunction with the NGWA/AGSE annual meeting.

One could conclude from these efforts and all of the activities surrounding pump and treat that this knowledge and information would now be incorporated into cleanups that are being planned and implemented in the field. Not! ("Wayne's World").

The author recently reviewed a remedial action plan for a Superfund site. Although the site was contaminated with nondegradable organics, the remedial action plan had described reaching "clean" as the remediation objective. Pump and treat was the only technology applied to the aquifer. Let us look at what this strategy cost.

Site Description

THE SITE CONSISTED OF THREE main contaminant sources. The sources were not contiguous, yet each was reported to contain similar contaminants. All of the contaminants were chlorinated hydrocarbons, and were nondegradable in a natural aquifer system. The contaminant plume was threatening a municipal well field. Blocking wells were installed several years ago to intercept the contaminants before they reached the municipal well field. The Superfund site encompassed all of the original sources of contamination, the blocking wells, and all of the area in between. At the same time that the blocking well system was installed, one of the original contaminant sites installed a pump and treat system to remove one of the sources of contaminants.

The stated objective of the remediation was to "clean" the entire site. The EPA proposed to use two more pump and treat and vapor extraction systems (VES) at the original sources of contamination, and a new blocking well system with a treatment system. The purpose of the new blocking wells was to ensure no contaminants penetrated the existing blocking wells and to reduce the amount of time the pump and treat system would take to remediate the area between the blocking well systems.

Cost Analysis

THE USEPA ESTIMATED THE COST of implementing the selected remedies for the well field site at approximately $15 million (Table 2.1). This estimate included the implementation of institutional controls, the installation of additional blocking wells and treatment system, the installation of source area groundwater extraction and treatment systems, and source area soil vapor extraction and treatment systems at each of the three contaminated sites.

The problem with this alternative is that pump and treat has not proved to be an effective method of reaching "clean" at contaminated Superfund sites as discussed previously. To reinforce this argument, the Superfund site was already treating one of the contaminated sites with pump and treat. The data from the first year of operation are summarized in Figure 2.1. The contaminants reached an asymptote after about 100 days of operation (Figure 2.1). The problem is that the asymptote is at approximately 2,500 ppb total volatile organic (TVO), and "clean" has been defined at 100 ppb TVO.

An alternative strategy was recommended to the USEPA (Table 2.2). This remedial alternative emphasized institutional controls to reduce exposure to contaminated media, optimization and sole use of the existing blocking wells to protect the well fields, and soil vapor extraction to eliminate migration of soil contaminants into source area groundwater. It also considered operation

TABLE 2.1. USEPA ALTERNATIVE

New Blocking Wells–Existing Blocking Wells–Source Area Soil Vapor and Groundwater Extraction and Treatment

Capital	Lump Sum
Mob./Demob., Site Prep., Health & Safety	$733,000
New Blocking Well System	886,000
Source Area Groundwater Extraction and Treatment	1,230,000
Site A Soil Vapor Extraction and Treatment	524,000
Site B Soil Vapor Extraction and Treatment	174,000
Engineering, Administration, and Contingencies	2,610,000
Total Capital Cost	**$6,160,000**

Operations and Maintenance (O&M)	Annual
Existing Blocking Wells	$164,000
New Blocking Wells	173,000
Source Area Groundwater Extraction and Treatment	218,000
Source Area Soil Vapor Extraction and Treatment	239,000
Groundwater Monitoring	47,000
Subtotal Annual O&M	841,000

Total O&M at Present Worth (5% over 30 years)	**$9,100,000**
Grand Total	**$15,260,000**

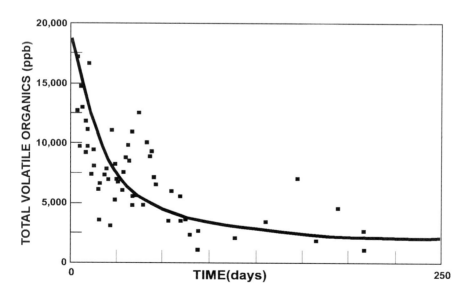

FIGURE 2.1. Total volatile organics versus time. Adapted from USEPA, 1989.

TABLE 2.2. RECOMMENDED ALTERNATIVE STRATEGY

Existing Blocking Wells–Source Area Soil Vapor Extraction and Treatment

Capital	Lump Sum
Mob./Demob., Site Prep., Health & Safety	$733,000
Site A Soil Vapor Extraction and Treatment	524,000
Site B Soil Vapor Extraction and Treatment	174,000
Engineering, Administration, and Contingencies	1,150,000
Total Capital Cost	$2,581,000
Operations and Maintenance (O&M)	**Annual**
Existing Blocking Wells	$164,000
Source Area Soil Vapor Extraction and Treatment	239,000
Groundwater Monitoring	47,000
Subtotal Annual O&M	450,000
Total O&M at Present Worth (5% over 30 years)	**$3,170,000**
Grand Total	**$5,751,000**

and maintenance for 30 years. The goal of this system was to protect human health and safety. The projected cost for this remedy was under $6 million.

At the present time, the EPA is forcing the PRPs to proceed with the original $15 million dollar program. All of the evidence from other Superfund sites and the specific data from the current site were not enough to persuade the EPA to change their strategy.

Strategy

THE BETTER STRATEGY AT THIS site would have been the alternative treatment program. The first step should have been to change from a "clean" objective to the protection of human health and the environment. It would be better to only spend the $6 million now, and put the other $9 million in a bank account. During the next two to five years, someone may develop a method that enhances the removal efficiency of pump and treat. The $9 million would then be available to implement that new technology.

Forming a strategy near the beginning of the project is important. As can be seen by this article, the first step is setting the right objective. There is significant money at stake when we ignore experience. One of the most clear experiences that we have, so far, is that all sites cannot be "clean."

REFERENCES

1. U.S. Environmental Protection Agency. 1989. Evaluation of Groundwater Extraction Remedies, Vol. 1, Summary Report. EPA/540/2-89/054, Washington, D.C.
2. Keely, J.F. Performance Evaluations for Pump and Treat Remediations. Ground Water Issue. U.S. Environmental Protection Agency, Office of Solid Waste and Emergency Response, EPA/540/4-89-005, October 1989.
3. Technical Exchange. 11/8/91, Quotation from draft memo issued by EPA Assistant Administrator Donald Clay.
4. *Federal Register,* Vol. 57, No. 31. Friday, February 14, 1992, pp. 5453–5454.
5. "Wayne's World."

3

WHERE IS THE MONEY?

Evan K. Nyer and Jeff Thomas

AFTER ALL OF THESE YEARS in the groundwater field I seem to be finally coming into fashion. Engineers are required for remediation, and everyone is doing remediation. Usually, the hydrogeologists begin the work at a site. They perform all the site characterization work, determine the location of wells, including the anticipated extraction rates, and then hand the project over to the engineers. The engineer is considered the key to the next stage in the process of remediating a contaminated site, and the clients feel that the engineer will be the one to save them money.

This perception is wrong. While the engineer will have a lot of work to do, the hydrogeologists are not free to return to the oil business. The price of oil is still too low, and there is a lot of hydrogeological work left to do during the remediation portion of the project.

I have stated many times that good remediation design requires a team of professionals in order to develop the most cost-effective remediation design. Let us go through a very simple example in order to illustrate this in one more way. We will use a simple gas station remediation over a 20 year period as our example. I have asked Jeff Thomas of Geraghty & Miller's Denver office to assist with this example.

Site Description

A SERVICE STATION HAS EXPERIENCED the release of unleaded gasoline from an underground storage tank (UST). An investigation has determined that the release was limited. No free product was determined to be present, though dissolved gasoline products were detected in several monitoring wells. The concentration of total petroleum hydrocarbon-gasoline in the groundwater beneath the site averaged 20 mg/L, while benzene was present at an average concentration of 2.0 mg/L. The geology under the site was characterized as a silty-sand to a depth of 30 feet. The silty-sand is underlain by a low permeability clay layer. Groundwater occurs at a depth of 14 feet below ground surface. Figure 3.1 provides a plan view of the site.

It was determined that a groundwater extraction rate of 10 gallons per minute (gpm) would be required to maintain hydraulic control of the plume area. The 10 gpm would also be sufficient to flush water through the contaminated area and remediate the site. (Note: This is old thinking.)

Six monitoring wells were installed during the investigation. It was determined to leave these monitoring wells in place in order to follow the progress of the remediation. These data were summarized and submitted to the state. The state reviewed and approved the report and requested that a remediation design report be prepared and submitted.

Remediation Design

A REMEDIATION SYSTEM WAS DESIGNED to treat 10 gpm of extracted groundwater. The remedial process design incorporated two extraction wells which discharge extracted groundwater into an equalization tank, followed by a single column air stripper for TPH-g and VOC removal prior to discharge into a city sewer. For the purposes of this evaluation, certain costs that were constant throughout all of the evaluation (e.g., project management and regulatory reports) were eliminated from the cost evaluation. However, we have tried to use enough real data to make the example plausible. (WARNING! The costs presented here are not real; it's sort of like that Murphy Brown thing.)

Table 3.1 summarizes the capital and operating costs from this system. The main operating costs from this type of design come from the operation of the treatment system itself and the monitoring of the site. Let us assume that the treatment system does not require significant operator attention or maintenance. We will use $1,000/month as the operating cost.

We will use EPA Methods 8015 and 8020 at $85/sample and $80/sample, respectively, to monitor the site. It will take $680 of manpower, equipment rental, and disposable sampling gear every time we go out for a sampling event. We will also assume two QA samples will be required for each sampling event. The standard monitoring interval on previous projects had been monthly.

The operating costs based upon these assumptions are summarized in Table 3.1. As can be seen, each sampling event at the site costs $2,000. The operation of the treatment system costs $1,000 per month. The total operating costs

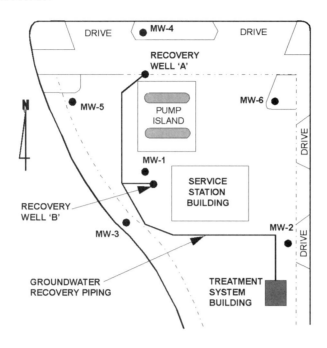

FIGURE 3.1. Plan view, gasoline station.

TABLE 3.1. CASE 1: PUMP AND TREAT; 6 MONITORING WELLS–MONTHLY SAMPLING

Capital	Units	$/Unit	Total $
Air Stripper Components	1	20,000	20,000
Air Stripper Installation	1	10,000	10,000
Monitoring Wells Installation	6	2,000	12,000
Extraction Wells Installation	2	2,000	12,000
Engineering Design			7,000
		Total Capital	$54,000
Operating			
Treatment System Operations		$1000/mo	$12,000/yr
Monitoring			
Labor & Equipment		680	680
Analytical	6(+2QA)	165	1,320
		Monthly Monitoring	$2,000
		Yearly Monitoring	$24,000/yr
		Operating Costs Year 1-20	$36,000/yr
		Cumulative Capital & Operating Costs - 20 yr	$774,000

of the remediation program, including monitoring, are $36,000 per year. The project is projected to take 20 years to complete remediation.

Alternative Remediation Design

THE OWNER OF THE GAS station was presented with a copy of a draft remediation report. He looked at the projected costs and realized that he would have to spend almost $800,000 over the 20 years to clean the site. And this station was just one of 50 gas stations that would require remediation during the next few years.

The owner decided that he needed "Advanced Technology" in order to get the costs lower. He called in an advanced engineering firm to review the remediation design. The firm reviewed the investigation report and the draft remedial design. They came up with the perfect answer: natural, biological in-situ remediation. From previous work that the firm had completed in other states, they felt that the site would clean itself naturally. There was no free product at the site. The ground surface was uncovered, and the soil was silty-sand, which should transfer oxygen from the surface. A reevaluation of the original data in the investigation report showed that the plume was not moving. This was strong confirmation that the natural biological reactions would be able to remediate the site. The best part was that this design would completely eliminate the treatment system and all of the corresponding O&M costs.

The owner was ecstatic and a meeting was scheduled with the state EPA in order to present this remediation plan. The state was also excited about the design. They had read several articles about the potential of biological treatment, but they had never used this process in the state. From their readings they had several questions. The main ones that could not be answered to their satisfaction at the meeting were:

1. Would the bacteria mobilize the gasoline products and allow them to migrate deeper into the aquifer?
2. Would the bacteria really prevent the horizontal and vertical migration of the plume?

The state regulators wanted to be reasonable, and they approved the remedial plan on the condition that the owner would collect extra data to show that the plume would not spread. They also wanted data that would show that the site would clean up in the same 20 year time frame as projected for the pump and treat. They requested that only two additional monitoring wells be added deep in the aquifer to show that there was no vertical migration of contaminants, and that only two monitoring wells be added downgradient of the plume to show that no migration was occurring horizontally. Finally, they requested that only two monitoring wells be added toward the center of the plume to show that the plume was actually shrinking and the site would clean up in 20 years. If the owner would comply with these requests, he could completely eliminate the treatment system as long as he maintained the monitoring schedule suggested in the draft remediation report.

The owner said that he would think about it and get back to the state. He went back to his office, had the advanced engineering firm send their report to the original consulting firm. A meeting was scheduled with all of the parties for a final review.

The original consultants brought Table 3.2 with them to the meeting. As can be seen in Table 3.2, the new design by the advanced engineering company did not produce the real savings that they anticipated. While the new design eliminated the treatment system and the O&M, the new design required additional monitoring. In fact, the new monitoring costs were equal to the total savings from eliminating the O&M from the treatment system. The only savings came from the capital cost of the treatment system, $30,000.

The owner was very disappointed. However, the hydrogeologist who did the original investigation had a suggestion. She suggested that they approach the state with the original design, but a reduced monitoring program. The new monitoring program would consist of monthly monitoring of the six monitoring wells for the first three months, quarterly monitoring of four monitoring wells for the next nine months, semiannual monitoring of three monitoring wells for the next two years, and annual monitoring of three monitoring wells for the last 17 years. She explained that the original investigation showed the plume was not moving, and the state was very familiar with pump and treat. From this experience base, they would be more confident of the progress of a pump and treat remediation. Less data would be needed to confirm this progress. They had been accepting reduced monitoring on other projects.

Table 3.3 shows the yearly cost of the new monitoring program. The capital and O&M from the treatment system remained the same and are shown in the first section of Table 3.1. However, the monitoring costs dropped dramatically from that original design. The total operating costs dropped to $23,000 the first year, $15,000/year for years 2 and 3, and dropped further to $13,500 per year for years 4 through 20. Figure 3.2 shows the cumulative costs from the three design cases. As can be seen, the reduced monitoring case provides over $400,000 in savings compared to either of the other designs.

The state was contacted and they agreed to the reduced monitoring plan. They were very comfortable with the pump and treat remediation. They had

TABLE 3-2. CASE 2: IN SITU TREATMENT; 12 MONITORING WELLS–MONTHLY SAMPLING

Capital	Units	$/Unit	Total $
Monitoring Wells	12	2,000	24,000
		Total Capital	$24,000
Operating			
Monitoring			
Labor & Equipment		690	690
Analytical	12(+2QA)	165	2,310
		Monthly Monitoring	$3,000
		Yearly Monitoring	$36,000/yr
		Operating Costs Years 1-20	$36,000/yr
		Cumulative Capital & Operating Costs - 20 yr	$744,000

TABLE 3.3. CASE 3: YEARLY MONITORING COSTS, PUMP & TREAT; 6 MONITORING WELLS–REDUCED MONITORING

	Monitoring Samples	Sample Frequency	Time	$	$/Yr
Year 1	6(+2) = 8	1/month	3 months	6000	
	4(+2) = 6	1/qtr	9 months	5010	$11,010
Years 2 & 3	3(+2) = 5	2/years	2 years	3010	$3010
Years 4–20	3(+2) = 5	1/year	17 years	1505	$1505

several systems operating in the state for many years. No problems had occurred with any of them. They felt that the monitoring schedule would confirm that the pumping system was controlling the plume and there was no need for the extra monitoring.

Conclusion

THE CLIENT ENDED UP VERY happy and the cost savings were due to the hydrogeologist (NGWA doubles my pay when the hydrogeologist is the hero). The true costs of a project are a combination of the treatment method and all of the operating costs including the monitoring costs. To get the most out of a project requires the combined expertise of engineers and hydrogeologists. If the project were more complex, the expertise of modelers, risk assessors, etc.,

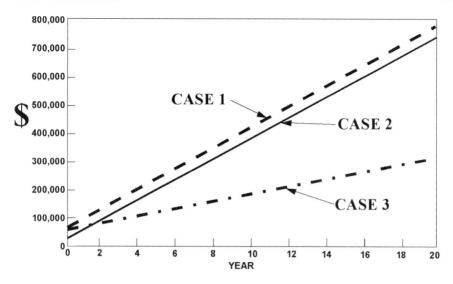

FIGURE 3.2. Cumulative costs.

might also be able to make a contribution. Remediation projects require teams of expertise.

The reader may feel that we have cooked the numbers to provide this example. I assure you all that the points made in this article are real. For example, I recently completed a Superfund design. The monitoring costs for a complicated design were still going to represent over one-third of the costs of the total remediation program.

A large portion of a remediation program is the cost of monitoring. We frequently agonize over the costs of the selected remediation process and its O&M costs, but overlook the costs of long-term monitoring for a site. The lesson here is that while we should be concerned about the costs for the remediation equipment, it may be very wise to spend more capital to remove contaminants as quickly as possible so that long-term monitoring can be reduced or eliminated. In very long remedial projects, the difference in capital cost of $10,000, $50,000 or $100,000 may become somewhat insignificant in view of the overall cost of the remedial program.

You are professionals, try these suggestions at home.

4

LIFE CYCLE DESIGN FOR IN-SITU REMEDIATIONS

Evan K. Nyer

THE LIFE CYCLE CONCEPT HELPS to focus the designer on the main strategies necessary to successfully remediate a site. The concept of Life Cycle Design and its use on groundwater was first published in 1985 in my book, *Groundwater Treatment Technology*. The simple basis for the Life Cycle concept is that groundwater remediations are unique, and that the requirements for the project will change over the life of the project. One must design for the entire life of the project, not just the conditions found at the beginning. Since 1985, I have used the concept of Life Cycle on groundwater treatment designs. Review of data from groundwater remediation systems during the past 10 years has provided a natural progression in our interpretation of the Life Cycle Design Concept. Originally, the Life Cycle concept provided a model by which a groundwater treatment system could be designed. In the late 1980s, we began to use the Life Cycle concept to indicate when a remediation system could be shut off. The current interpretation of the Life Cycle concept is that each remediation consists of two separate phases: "Mass Removal" and "Reaching Clean." Each of the progressive interpretations of the Life Cycle concept are discussed below. The current interpretation should be used on all future remediation designs and it will help the reader understand the results from past installations.

In 1985 the most common groundwater treatment technology was pump and treat. There was very little discussion on remediation methods. The main discussion was on the type of technology used to remove the organics and metals from the water withdrawn as the result of the pump and treat system. The first use of the Life Cycle curve was to provide a model that could be used to design the groundwater treatment system.

In the late 1980s many pumping systems had been running for several years. A couple of things started to become obvious. First, we were right about the concentration decreasing as the project progressed. The concentration in the pumping wells and in the aquifer itself decreased as the pump and treat system operated. Second, the concentration curve flattened, and the concentration stopped decreasing after the systems had been running for extended periods of time. This caused two problems: (1) The pumping systems were no longer removing significant amounts of contaminants; however, operational costs generally remained the same; and (2) The concentrations were not low enough to declare the aquifer clean and shut off the treatment system. The Life Cycle curve was once again used to describe the change, or lack of change, in the concentration. In addition, the Life Cycle was used to indicate when a treatment system should be turned off, but progress would still continue in the remediation of the aquifer. While the Life Cycle curve was still mainly concerned with aboveground pump and treat systems, our better understanding of in-situ processes started to affect our interpretation of the curve.

The current interpretation of the Life Cycle curve completes the incorporation of the in-situ reactions into the planning of a remediation. Once we accept that most of the in-situ technologies are based upon the use of water or air as the carrier, then the life cycle presentation of the change in concentration over the life of the remediation is accurate. Figure 4.1 represents the change in concentrations over the life of a pump and treat project. It also describes the life of a Vapor Extraction System, Sparging, or Vacuum Enhanced Recovery project.

Air has many advantages over water as the carrier in a remediation. Air will be able to remove the organic compounds from the ground at a much faster rate than techniques that use water as the carrier. The simple advantage of an air-carrier-based remediation being able to move more pore volumes than a water-based remediation in a given amount of time would shorten the amount of time required to remove an organic compound. Assuming that the comparison is based upon two projects that had the same amount of organic originally in the ground, the concentration would decrease much faster with air as the carrier, compared to water. The left side of the curve in Figure 4.1 would have a steeper slope. (Of course, if we used pore volumes for the x-axis, then the curves would be the same for water or air.) If the organic compound is highly volatile, then the curve would be even steeper with air as the carrier.

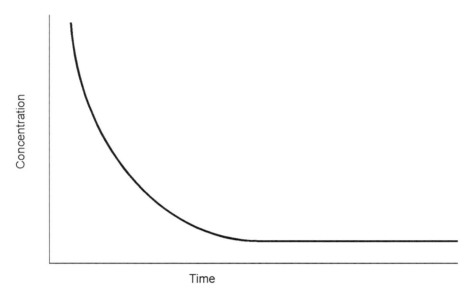

FIGURE 4.1. Life Cycle concentration during remediation.

However, the right side of the curve would not automatically be affected with a switch from water to air as the carrier. The right side of the curve is controlled by diffusion and the geology of the site. Air has many of the same limitations as water. The material that is not in direct contact with the main flow path of the carrier must diffuse over to the main path in order to be removed from the ground. The right side of the curve would look the same for water or air. So while the slopes would change, the overall curve would still remain an accurate description of the change in concentration over the life of the project.

The Life Cycle curve has always been an important tool in the design of a remediation system. This is still true for in-situ technologies that rely on one of the carriers. When we analyze the change in the shape of the Life Cycle curve between the two carriers and the various enhancements to the carriers, we realize that the left and right sides of the curve must be treated separately.

Figure 4.2 is now the best way to think of the Life Cycle curve when designing a remediation. While the curve is the same, we have completely separated the left half and the right half of the Life Cycle line. The left half represents the Mass Removal portion of the remediation. The right half represents the Reaching Clean portion of the remediation.

There are really two separate projects in any remediation. The first project is the removal of the maximum amount of mass of contaminants. The in-situ

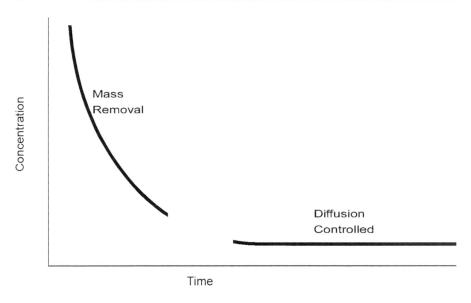

FIGURE 4.2. The best way to think of the remediation Life Cycle curve.

technologies that rely on air as the carrier have been shown to have a significant advantage in this type of project. (Sims [Sims, 1990] reported a factor of
10,000.) Most of the published information that we are currently receiving on
in-situ processes is really an analysis of the mass removal capability of these
techniques. The second project of any remediation is reaching the mandated
concentration required to declare the site 'clean.'

Pump and treat never failed as a technology for the mass removal type of
project. Water was capable of removing a significant portion of the mass of
contaminants from the aquifer. Air may cost less and remove the compounds
faster because of its advantages, but that doesn't mean that water failed. The
EPA and the NRC reviews were based upon the Reaching Clean part of the
remediation. When they stated that pump and treat was not capable of
remediating an aquifer, they really meant that pump and treat failed to reach
the concentration that would allow the aquifer to be designated as 'clean.'
Pump and treat failed in the Reaching Clean project. (Both reports made it
very clear that they felt pump and treat was a successful technique for controlling plume movement.)

The real problem with pump and treat systems is that the designers expected that the systems would be able to clean the aquifer, and they told everyone that. They did not include the geology and diffusion limitations of
using a carrier when they designed their remediation. Because of the slow
pace of pore volume exchange with water as the carrier, the results did not

show up for many years. This delay allowed thousands of remediation systems to be installed based upon the assumption that pump and treat could reach clean.

Air as a carrier has the same geology and diffusion limitations. However, we are now installing the in-situ technologies based upon air and expecting them to reach clean. We are taking the results from the first part of the remediation and declaring that since the compounds are being removed much faster, then obviously the site will reach clean.

This is why it is very important to separate the two projects. Mass Removal is completely separate from Reaching Clean. A remediation design should be analyzed as two separate projects. What will the technique do for Mass Removal? What will the technique do for Reaching Clean? Separating the remediation into two separate projects does two major things for the designer. First, we are able to predict at the beginning of the project what will happen over the entire life of the remediation. This not only gives us our design basis, but also provides the understanding necessary to interpret the data during the operation of the remediation. Second, we are able to realize that different technologies may have to be applied to the site in order to complete the remediation. One technology may have to be used for the Mass Removal, and a different technology used to Reach Clean.

Figure 4.2 is the representation of the Life Cycle of the remediation that we should be using for in-situ projects. It is still important to use the Life Cycle curve when developing a strategy for a remediation. Figure 4.2 incorporates our latest understanding of the different factors that affect the entire remediation project.

REFERENCE

1. Sims, R.C. "Soil Remediation Techniques at Uncontrolled Hazardous Waste Sites, A Critical Review," *Air & Waste Management Association*, May 1990.

5

DEVELOPING A HEALTHY DISRESPECT FOR NUMBERS

Evan K. Nyer, Terry Regan, and Deepak Nautiyal

IT IS TOO EASY TO read a number from a report and think that it is gospel. Reading, writing, and interpreting the term 5 parts per million (ppm) should not be a religious experience—it needs to be a scientific endeavor. The number 5 ppm really represents our abilities in sampling and analysis, the relationship of the properties of the compound to the biogeochemical conditions found at the site, and the instant in time at which the sample was taken.

There are seven factors that can dilute the meaningfulness of analytical data from soil and groundwater samples collected during environmental investigations or remediation. These factors must be considered when using analytical data to describe, evaluate, or regulate environmental conditions at a given site. To keep the theme going, Moses did not write down the number 5 ppm; some hardworking geologist, chemist or engineer probably did. Let's see if we can help everyone develop a little healthy disrespect for such a number.

SAMPLING METHODS

THE FIRST SET OF PROBLEMS encountered during site investigation results from the method(s) used to acquire the samples and from where the samples are analyzed.

Field Analysis, Laboratory Analysis, or In-Situ Measurement

IT IS IMPORTANT TO KNOW where groundwater samples are collected and analyzed. Some parameters, such as pH, temperature, and specific conductivity should be measured in the field because they are relatively unstable. For example, all else being equal, field pH data are more reliable than laboratory pH data because the pH of the sample is dependent upon temperature and other environmental factors. The pH of samples shipped to the laboratory will change in transit. Dissolved oxygen (DO) and redox potential (eH) are also very important analytical parameters, and, like the other parameters listed above, they should be measured in the field. Also, DO and redox samples are better measured in situ, and not after the sample has been removed from the well, since DO and redox concentrations can change when the sample comes in contact with the atmosphere. My favorite example with regard to analytical location is a report that summarized temperature data which were measured at the laboratory.

Volatile Organic Compounds

EVEN IN AREAS WHERE WE all recognize that care must be taken when sampling, several factors should be considered, including the following:

- *How volatile are the individual compounds?* Relatively speaking, in a given sample, the reported concentrations of the less volatile species may be higher compared to the reported concentrations of the more volatile species. Try pouring a sample with vinyl chloride from one beaker to another and see if anything is left.
- *Have soil samples been composited?* Mixing of the soil in the field will aerate the soil and result in loss of volatiles from the sample. This problem must be weighed against the cost of analyzing multiple soil samples.
- *What are the surface and subsurface conditions at the collection point?* Under buildings or paved surfaces, and beneath confining subsurface clay and/or silt layers, concentrations of the more volatile compounds are likely to be higher relative to similar subsurface conditions under an open field, where the more volatile compounds may naturally vaporize into the atmosphere.

Filtered vs. Nonfiltered Samples

ARE GROUNDWATER SAMPLES SUPPOSED TO represent the soluble components of the aquifer? Most regulatory agencies require unfiltered samples. New "low flow" sampling techniques are supposed to minimize solids in the sample.

The important point to remember is that to have a full understanding of contaminants and how to remediate the aquifer, one has to know if we are dealing with a soluble or solid phase of a compound.

Most metals, with the most notable exceptions of hexavalent chromium (Cr^{+6}) and arsenic (As), usually are not found dissolved in groundwater under normal subsurface conditions of pH and temperature. Rather, they occur in the subsurface adsorbed onto soil particles or as precipitates. However, groundwater samples for metals analyses are often preserved in the field with acid to ensure that the metals do not precipitate out of solution during transit. The acidification will desorb the metals from suspended soil particles in the sample. It is important for the investigator to know whether the compound is solid or soluble. Although filtering of samples may not be acceptable for regulatory or reporting purposes, it is a valid economical means to collect important data for remediation and treatability purposes.

Like metals, many organic compounds have a very low solubility, and are, therefore, not found in the dissolved phase in groundwater samples. For example, polychlorinated biphenyl (PCB) and other high-molecular-weight organic compounds detected in groundwater are an indication that there are suspended particles or oil droplets in the sample.

ANALYTICAL STATISTICAL VARIATION AND OTHER ERRORS

At best, the concentrations of constituents in an environmental sample reported by an analytical laboratory are only a snapshot of conditions which existed at the precise location and time that the sample was collected. In actuality, it is very difficult to duplicate analytical results in groundwater, and more so in soil. For example, if multiple samples were collected simultaneously from the same location, you would likely get different analytical results for each sample. The results would follow a normal distribution (Figure 5.1). The problem is that if only a single sample is taken, there is no way to determine where the analytical result falls on the normal distribution. It cannot be determined based on a single analytical result, whether that value is in the high or low range, or closer to the average. In other words, would a 5 ppm result be part of a sample pattern that represents a 6 ppm normalized concentration?

Another common problem with laboratory analytical data occurs when concentrations in the sample are at or near the analytical detection limits. It is important to remember that concentrations should not be reported as "not detected," but rather "below the detection limit." This becomes very important as technology devises more sophisticated instruments, which for some

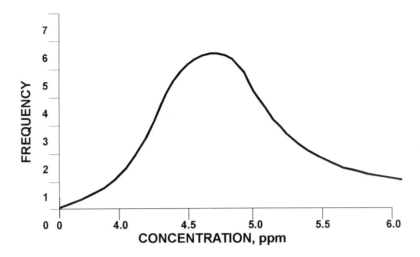

FIGURE 5.1. Normal statistical variation in analysis.

analytes, can already detect concentrations in the parts per quadrillion range. Also, when other compounds are present, the detection limit can be altered significantly based on the concentrations of the other components, and on dilution factors.

One final problem with laboratory and other analytical techniques that require instrumentation is calibration. All measuring devices require the operator to tune the device against a standard. If a mistake is made during the process (e.g., wrong standard compound; inappropriate concentration range), then all results from the instrument after that calibration are inaccurate. Don't assume just because the number pops up in digital form, that it is right.

GRAPHICAL DATA PRESENTATION/INTERPRETATION

A COMMON TOOL USED TO delineate concentrations in the subsurface is to prepare a contour map of an area of concern. The contours are drawn by connecting points of equal concentration, and interpolating between data points. The danger arises when there are sparse data points to contour. The contouring then becomes a much more subjective exercise which can cause problems.

For example, there was a site where two distinct sources of contaminants were mistakenly merged and shown as a single contaminant plume because the investigator closed contours surrounding both source areas even though there were insufficient data to make such a speculative interpretation. Figure 5.2 shows the contour lines interpreted by the original investigation, assum-

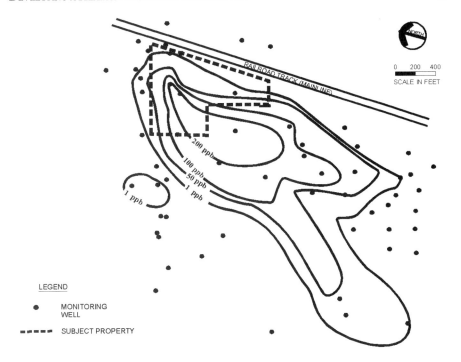

FIGURE 5.2. 1,1-Dichloroethene distribution. Single sources.

ing a single source. Figure 5.3 shows contour lines based upon two sources. Once contours are put onto a map, they are taken as fact, and trying to convince someone otherwise is like trying to put the toothpaste back into the tube. Obviously this can have serious implications at sites where litigation over responsibility is ongoing. To avoid misinterpretation of contoured concentration maps, the following rules should be applied:

1. Always show the control points. The density of the control points is an indication of the confidence in the contouring.
2. Use dashed contours liberally in areas where the data are interpolated among sparse control points. This indicates that the contours are inferred, and there is a lower confidence level in the contouring.
3. Make allowances for known surface and subsurface features that may have an influence on your interpretation, such as hydrogeologic conditions, building foundations, or streams. Ask yourself if the interpretation makes sense.
4. As with any map, you must include a north arrow and a scale; and, even more importantly, with concentration data you must also always date the map, because concentrations change in the subsurface over time.

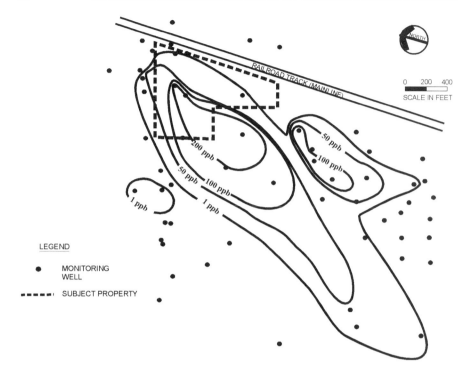

FIGURE 5.3. 1,1-Dichloroethene distribution. Two sources.

CHEMISTRY

EVEN IF YOUR DATA EVALUATION/validation determines that the reported 5 ppm analytical result is correct, you still have to understand what that number tells you about conditions at the site. Solubility and retardation are two parameters which for various compounds can produce different interpretations of the same concentration.

Solubility

THE SOLUBILITY OF DIFFERENT CHEMICALS in groundwater varies greatly. Therefore, the same concentration of different chemicals can mean very different things. For example, if 5 ppm of trichloroethene (TCE) were detected in groundwater, there would be serious concern that the source of this concentration was a nearby pool of dense nonaqueous phase liquid (DNAPL). This is because TCE is not very soluble in water, and concentrations as high as 5 ppm are indicative that pure TCE exists in the subsurface nearby. Remediation of DNAPLs from the subsurface has proved to be very difficult, expensive, and in many cases, impossible. On the other hand, 5 ppm of acetone in groundwater would not have the same significance as 5 ppm of TCE, because acetone is completely miscible in water, and very large concentrations of acetone can dissolve into, and move along with groundwater flow. For this example, the remediation issue is compounded by the fact that TCE has a relatively high toxicity compared to acetone, resulting in much lower cleanup standards for TCE than for acetone.

Retardation

RETARDATION IS THE TERM USED to describe the rate of contaminant transport in groundwater relative to the rate of groundwater flow. Some contaminants are slowed down by interactions with the soil particles through which they are flowing, depending on such factors as organic content and clay content of the soil. For example, the movement of chloride will not be retarded relative to groundwater flow, but the movement of TCE will, because of the interactions of TCE with the soil.

Compounds that degrade are also retarded relative to groundwater flow rates. Therefore, the investigator must consider whether the compound is not at a location because it has been destroyed and will never reach that location, or it is simply moving slowly because of interaction with the soil particles.

GEOLOGY

THE GEOLOGY OF THE SUBSURFACE is relevant when evaluating analytical data, especially as it relates to how the data are to be used. This is true on a macroscopic scale and also on a microscopic scale. On a macroscopic scale, the aquifer characteristics must be understood. In the simplest case of the often-cited-but-rarely-seen homogeneous isotropic aquifer, the evaluation of the analytical data can be relatively straightforward, because many of the assumptions used to describe and understand groundwater flow systems can be applied with confidence. This is true whether the objective is to investigate/assess subsurface conditions, or to collect data for designing remedial solutions.

As the subsurface geology moves away from the ideal textbook case toward the more complex, anisotropic and nonhomogeneous conditions, the assumptions break down and the meaning of the analytical data must be viewed with a lesser degree of confidence. In more complex systems such as those with layered or discontinuous stratigraphy or fractured bedrock, an understanding of the three-dimensional flow field becomes more important. The cost of subsurface investigations generally increases proportionately with the complexity of the subsurface conditions, because more data are needed to characterize and understand the three-dimensional flow field.

Understanding geology and groundwater flow on a microscopic level may also be required when evaluating analytical data. At the microscopic level, the mechanisms of groundwater flow through individual pore spaces within the geologic matrix must be understood in order to evaluate analytical data relative to the data-collection objectives. For example, if the objective is to describe plume migration, it is important to understand that advection is the main mechanism by which contaminants move along the major flow paths within the pore spaces. However, within this microscopic flow system, diffusion and dispersion may be important factors in the movement of contaminants in the subsurface as well. This information must be considered if the objective of the data collection is to design a remedial system to remove contaminants from the subsurface.

BIOGEOCHEMISTRY

INTERACTIONS BETWEEN CONTAMINANTS AND THE subsurface and natural biodegradation processes can significantly impact calculation of the mass of contaminants present in a given source area. As will be discussed below, this, in time, could then affect interpretation of a site and the remedial design.

Mass Balance Approach

SOIL AND GROUNDWATER CONTAMINATED SITE cleanup efforts are commonly planned and designed based on inaccurate estimation of the quantity of contaminant mass to be addressed. It is ironic that while a mass balance is routinely conducted on aboveground treatment processes, a mass balance approach is generally not applied to the subsurface contamination system. This is because it is generally extremely difficult to attribute the percentage of mass in the subsurface that is (1) persistent, (2) mobile, and (3) chemically and/or biologically degraded with time. One of the most difficult determinations is the total mass of NAPL presence. The most common data used to determine NAPL presence below the groundwater surface are concentrations of the compounds in the groundwater. There is no way to tell the difference between a small amount of NAPL mass dispersed in the aquifer and a large amount of NAPL in a single location unless you are lucky enough for one of the boreholes to hit the mass directly. However, it is very important to calculate mass balance for subsurface systems at a site because it inherently addresses site characterization and remediation. The mass balance approach should focus on defining the following:

1. the location and form of contamination;
2. the chemicals at the site in context of mobility versus degradation;
3. the treatment approaches based on specific waste phases at specific times during remediation; and
4. a monitoring program based on the information about specific chemicals in the specific phases in the subsurface at specific times.

Interactions Between Contaminants and the Subsurface

IN EVALUATING ANALYTICAL DATA, IT is extremely important to understand the actions between specific contaminants and site-specific subsurface properties. This allows the investigator to weigh the significance of the analytical data with respect to contamination of various media (soil/groundwater/air) due to mobility and/or persistence of chemicals at a site. Subsurface soil at various sites may be heterogeneous due to factors such as mineral content of soil, size of soil particles (clay vs. sand or silt), and organic carbon content (pineland vs. marshland). Heterogeneity will partially determine how a contaminant would reside in the soil matrix, so it is critical to look at these properties while assessing mass transport in the subsurface. For example, the greater the total organic carbon (TOC) content of the soil, the greater the absorptive capacity for volatiles and, therefore, the less mobile the volatiles become.

The behavior of hexavalent chromium (Cr^{+6}) vs. trivalent chromium (Cr^{+3}) provides another example of how subsurface conditions interact with contaminants. Under normal subsurface conditions, Cr^{+3} will bond to the soil and be immobile. On the other hand, Cr^{+6} is very soluble in water and has very little interaction with the soil particles. Because Cr^{+6} is more mobile (and also more toxic) than Cr^{+3}, it is desirable to get the chromium in the Cr^{+3} phase. This can be accomplished by adding chemicals (e.g., sugar) to the subsurface to maintain reducing subsurface conditions. Therefore, when evaluating chromium analytical data, the investigator must also have an understanding of the pH and oxidation/reduction potential (ORP) of the subsurface to accurately evaluate what the data truly represent.

Biodegradation Process

SINCE BIODEGRADATION PROCESSES MAY PLAY a key role in the overall cleanup process at any contaminated site, it is very important to assess the potential for biological degradation when looking at analytical data. The assessment process should be based on material balances and mineralization approaches to determine the environmental fate and behavior of the contaminants in the site-specific soil. The rate of biodegradation can be determined by measuring the loss of parent compound and production of carbon dioxide with time, as well as formation and disappearance of intermediate by-products of the bio-degradation processes. Therefore, the investigator should review the analytical data for the present of the contaminants as well as the by-product compounds. For example, the degradation pathway of TCE, PCE, and 1,1,1-TCA results in the production of six, seven, and four by-products, respectively. Vinyl chloride, a common by-product of these compounds and a carcinogen, is highly persistent under anaerobic conditions, but degrades under aerobic conditions. Therefore, when designing a bioremediation system for these compounds, emphasis should be placed on ensuring total detoxification as well as disappearance of the parent compound. For halogenated compounds, a full description of the geochemical conditions would be required in order to understand the significance of a 5 ppm analytical result for any of the compounds.

TIME

S--T HAPPENS. A 5 PPM analytical result only represents the concentration at the point in time at which the sample was taken. Analytical data review should also incorporate a comparison of the rate of transport with the degradation rate for each chemical of concern to determine if the transport is significant relative to degradation. For example, if the source is removed, benzene, which is highly degradable, will usually degrade and not move more than a fixed distance from the source (i.e., the degradation rate will be faster than the transport rate). More than one sample is required over a significant period of time in order to truly understand what one 5 ppm analytical result really means.

Biodegradation is not the only process that is active at contaminated sites. At one site the organic compound was not degradable, but it still only had a half-life of six years. Other chemical processes (e.g., hydrolysis) combined to destroy the compound. Only a time evaluation led to the understanding that other destructive factors were important at this site. The investigator must consider properties like half-lives because if a 5 ppm concentration is only going to be 2.5 ppm in six years, the entire approach to investigation, understanding, and remediation will be changed. We will not be able to collect enough information to fully understand and anticipate all of the processes that can affect a single 5 ppm concentration of a specific compound. Time itself must be sampled and analyzed.

CONCLUSION

THAT FIVE INCH THICK REPORT that cost several millions of dollars to produce only weighs as much as the tablets upon which the ten commandments were written. It doesn't carry the same weight of meaning. The numbers in that report are just ink ruining perfectly good white paper unless the person using those numbers understands all of the variations that are inherent in the sampling, analysis, interpretation, and presentation.

OK, all together let's chant, "We Don't Believe."

6

PRACTICAL REMEDIATION TECHNOLOGIES

Evan K. Nyer

INTRODUCTION

As we all know, there is going to be a test next week. It will be 50% of your grade, and therefore an important part of your passing this course. In order to give everyone the best chance of success, the purpose of this column will be to review the main points that we have covered. (Everybody take two breaths and get over the flashback. Aren't mind games fun?)

Seriously, a little review never hurt anyone. While we have discussed many different specific topics recently, there have been two major themes that we could place most articles in. First is that geology controls most remediations. We have discussed this as far as reaching remediation, design of remediation techniques, and strategies for selecting the right combination of remediation technologies. The second major theme has been the separation of remediation technologies into two major categories. The first category is the old pump and treat in which we used water as a carrier. The second category, and newer technologies, are all related in that they use air as the carrier.

Now I realize that these are simple ideas and that most of you understand them and are now bored with my discussing them. However, there is a significant portion of the readers who only concentrate on the flashy columns that have described the application of new technologies. As I have stated many times before, the only way to design a cost-effective remediation is to understand these two basic areas, and the only way you're going to pass this test will be to understand these two areas. The flashy new technologies will only count as 10% of this test. So, let us review both basic ideas one more time.

GEOLOGY CONTROLS REMEDIATION, NOT TECHNOLOGY

NO MATTER HOW NEW OR fancy the technology is, its application will have limits when applied in the field to remediate a site. Those limits are related to basic geologic formations and properties. The limits show up in two important ways. First, we are finding that most contaminated sites will not reach MCL concentration for the particular organic while using water as a carrier. The second is that all technologies based on a carrier, water or air, will have a change in concentration over the life of the remediation that can be called the Life Cycle.

The basis of all pump and treat remediation systems is that water is used as the carrier to bring contaminants from the aquifer to aboveground, where the contaminants are removed from the water. There have been two prominent studies performed during the last several years that focused on the limitations of pump and treat as a remediation technology for aquifers. The EPA and the NRC both studied multiple sites at which pump and treat was applied to determine the capabilities of this technology. The details of these studies are available in other published work (Nyer et al., 1996; Nyer, 1995).

The good news from pump and treat systems is that two functions are carried on at the same time with a pump and treat design. The first function is to use the removal of the water from the aquifer to create a new flow pattern in the aquifer. This flow pattern, if designed correctly, can change the direction of groundwater movement. Since all plumes move by advection, controlling the water movement controls the spread of the plume. Both the EPA and NRC studies concluded that the control portion of the remedial design worked very well at most sites for pump and treat designs.

The second function of the pump and treat designs is to return the aquifer to "clean." Clean used to be defined mainly by the MCLs of the individual organic compounds. Both studies found that the pump and treat designs had severe limitations when trying to bring an aquifer back to clean conditions. In general, sites have an initial period of rapid removal of mass from the site, followed by gradual lessening of the rate of mass removal, ending in a long period of constant mass removal. The data pattern is consistent when shown as concentrations. The consistent mass removal rate or concentration can be described as the asymptote of removal curve.

This changing concentration over the life of the project is shown in Figure 6.1. Figure 6.1 is the Life Cycle of the concentration of the remediation sys-

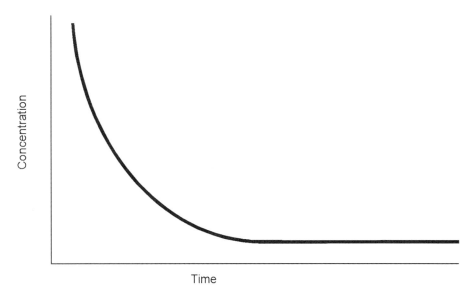

FIGURE 6.1. Life cycle concentration during a remediation project.

tem based upon pump and treat. While the scale on this curve may change depending on the geology and the organics involved with each individual project, the general shape of the curve occurs on every remediation project. What both studies found was that this consistent concentration was normally above the concentrations required by the MCLs. Some of the monitoring wells would reach clean, but the entire site would not return to precontaminated conditions. In general, the wells near the original source of contamination will not go below MCLs. According to the NCR, this was true for all sites that had chlorinated organics, and most sites that had degradable organics.

The Life Cycle curve can teach us several things about good remediation design. The first thing is that the design of all aboveground treatment equipment must anticipate the change in concentrations over time. While we have only been discussing water treatment equipment so far, this statement is also true for treatment designs based upon air as a carrier. The aboveground treatment equipment associated with Vapor Extraction Systems, Vacuum Enhanced Recovery, and Air Sparging all have to be designed with the change in concentration in mind. This is because geology controls the shape of the Life Cycle, not the technology or the carrier that the technology uses.

The next lesson that the curve can teach us is that we can only design for the front part of the curve. Good technology and design can change the steepness of the drop in concentration at a site or, more importantly, the rate of mass removal from a site. The lower portion of the curve, the asymptote, cannot be controlled by technology, however. This portion of the curve is controlled by diffusion in the aquifer. Figure 6.2 shows a representation of a standard homogeneous aquifer. Even when the geologic conditions create homogeneous characteristics in the aquifer, certain flow patterns are set up in the ground for carrier movement. The simplest way to explain movement of carriers below ground is that there are main flow patterns of carrier movement, minor flow patterns, and relatively stagnant areas. The lower part of the curve is controlled by the diffusion from the relatively stagnant areas of the aquifer into the minor and main flow patterns. Details of these flow patterns and the limitations that they cause for remediations are provided in other references (Nyer et al., 1996).

The bottom line is that any remediation design that has an objective of removing mass from the aquifer must be designed for the initial part of the Life Cycle. The cost-effectiveness of this design will be based upon the ability to remove the equipment and stop operation. If the equipment is left in place for the entire Life Cycle, the advantages of the technologies will not be realized in cost savings to the project. Unless that equipment is shut down and removed after the mass removal portion of the project has been completed, no savings will occur. This will be discussed further in the next section.

The final area that must be considered in all Life Cycle designs is monitoring. This is one area that most designers do not feel is important in their original design. However, due to the length of time most remediation systems and full cleanups require, monitoring can become a significant portion of the cost, especially in later years of the project. Therefore, the original design of remediation must include consideration of monitoring over the Life Cycle of the project. As remediation progresses and the contaminants are brought under control and removed, the number of monitoring wells used, the frequency of their use, and individual chemicals monitored must reflect the conditions of the site over time. In other words, the monitoring must be wound down over the life of the project in order for a cost-effective remediation to occur.

AIR VERSUS WATER REMEDIATION TECHNIQUES

THE OTHER MAIN THING THAT we have kept constant over the last several years is that there is no such thing as in-situ technologies. Through several examples we have shown that the main three in-situ technologies—vapor extraction systems, sparging, and vacuum enhance recovery—all still rely on a carrier to remove the compounds from the aquifer. All of these methods are based on air as the carrier, as opposed to water being the carrier. Due to the nature of air as a carrier, the air carrier methods can have an

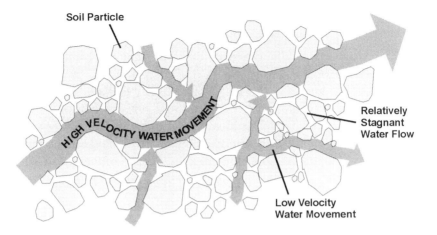

FIGURE 6.2. Water flow through soil microcosm.

advantage of 200 to 10,000 over water as a carrier, depending on the nature of the organics and the geologic conditions. We normally think of pump and treat methods over a 10 to 20 year period, while we think of air methods over a 6-month to 2-year period.

Advantages of these methods are significant but they only address the initial part of the Life Cycle curve. As with water as a carrier, the air will have areas of main flow patterns, areas of minor flow patterns, and areas of relatively stagnant conditions. Once again, these relatively stagnant conditions will control the lower part of the Life Cycle curve and produce the asymptote of low concentration. This asymptote is controlled by the geologic conditions of the site. This relates right back to our original comment that geology controls remediations, not technology. This statement is true in the air methods in addition to the water-based carrier methods.

The key to good air-based design then becomes the length of the time the equipment is on site. Since water and air both hit asymptotes, it is important to separate the conditions at which the equipment is removed from the site from the conditions that signify the site is "clean." If the asymptote (or rebound) is above the MCLs, and the end of the project is defined as MCLs, then an air-based method of remediation can be required to be on site for many years. Most of the savings between an air sparging system and a pump and treat system are lost if the sparging system has to run for 25 or 30 years. The basic advantage of air system is mass removal. The monetary advantage is that this mass removal from the air system can happen in a relatively short period of time of 6 months to 2 or 3 years. The equipment must be removed if this advantage is to be turned into cost savings. The lower part of the Life

Cycle curve must then be dealt with by another remediation technology. We have discussed several biological, both microbiological and macrobiological (phytoremediation), technologies in conjunction with finishing the remediation project as reaches to the asymptote.

Cost-effective remediation must understand the advantages of air over water as a carrier. To fully realize the cost advantages, however, the designer must also realize the limitation of air as a carrier. Both the advantages and the limitations are necessary in designing a cost-effective remediation design.

CONCLUSIONS

SORRY ABOUT THE BORING REVIEW of "simple" stuff. The problem is that most of the design mistakes that I see in the field can be related to one of the basic ideas that I presented above. It seems that designers do not feel that they are doing a good job unless they put in the latest, most technologically advanced designs that they can find in a magazine article. This is true all over the world. The only time that I get a lot of requests for a copy of one of these columns from foreign readers is when a 'hot' technology is mentioned in the title.

Next column, I promise to return to ADVANCED technology. One possible subject could come from a project that I have just finished. This remediation took one year from the time we received verbal approval from the regulatory agency. In that one year we designed, installed, and operated a remediation system that covered an area of approximately 5 acres; required 38 well points screened 65 ft below land surface, two miles of interconnecting pipe, a water treatment system, and an air treatment system. In other words, it was not a small project. The system removed over 95% of the chlorinated organics from the aquifer in that time. Would the readers consider that advanced technology? The problem is that the design was not based upon any of the new or advanced technologies. It was based upon the basic criteria that I have presented in this column. It may be boring, but you can use these ideas to save significant amounts of money and time.

REFERENCES

Nyer, E. et al. *In-Situ Treatment Technology*, Geraghty & Miller Environmental Science and Engineers Series, CRC Press, Inc., Lewis Publishers, Boca Raton, FL, 1996.

Nyer, E. Life Cycle Design for In-Situ Remediations, *Groundwater Monitoring and Remediation,* Treatment Technology Column, Spring 1995.

EPA. Evaluation of Ground-Water Extraction Remedies: Phase II. Volume I Summary Report, Publication 9355.4-05. Washington, DC, U.S. Environmental Protection Agency, Office of Emergency and Remedial Response, 1992.

National Research Council. *Alternatives for Ground Water Cleanup,* National Academy Press, Washington, DC, 1994.

7

MONITORING

Evan K. Nyer and Lynne Stauss

I KNOW THAT ALL OF you would like me to use this space to discuss some hot new technology. I hate to disappoint all of you, but I am, once again, going to concentrate on a way to save money instead. You will not be reading about Air Sparging with Chlorine Gas (just kidding; this is not a real technology, and for those of you who do not believe me, consider this the first place of publication, and the patent is all MINE!). Today, I am going to discuss (preach about) monitoring.

I know, it sounds very dull. However, let us look at some numbers. Assume that there are 200,000 contaminated sites across the country that are being monitored. Each site has an average of 10 monitoring wells, and each monitoring well is sampled an average of once per quarter. Each monitoring event (sampling, analysis, and report) costs an average of $2500 per well. Now, this article is going to recommend that all of these sites switch to once per year monitoring. Finally, assume that the article is only 25% successful. In other words, I only succeed in eliminating an average of one sampling event for every well each year. The total cost savings would be $5 billion per year. As I have stated before, I will be glad to work on a small percent of savings basis.

I cannot think of any single technology (or all of them combined) that has the potential of saving multiple billions of dollars per year. While we have had a lot of fun doing investigations and remedial designs and installations, it is now time to start concentrating on the next big cost item in the process. In order for us all to do our jobs correctly, we are going to have to work on our monitoring plans.

Now, this is not going to be a simple or easy project. Just because I say we should switch to once per year monitoring does not mean that everyone is going to agree with me. My children are not the only people who do not take my advice. Also, this is too big for me to do by myself. Therefore, I have assignments. These assignments are broken into categories based upon technical background and career. They include work for geologists, researchers, regulators, and project managers. The assignments for each group are provided below. The project manager assignment includes suggestions on ways to reduce monitoring events.

I have asked Lynne Stauss from the Navy to assist me in writing this article. I have seen some of the initial work that the Department of Defense is doing in the area of monitoring and have been very impressed. Lynne's personal experience will be helpful for the reader, but because she is a government employee, all of you have to read the following statement before you are allowed to learn anything, "The opinions expressed in this article are solely those of the authors and do not represent official Department of the Navy or Department of the Defense positions."

Now that you all think that this paper is important, let us get on with the specific assignments.

GEOLOGISTS: Get back into the project and design a reasonable monitoring program.

The only member of the remediation team who can design a monitoring program is the geologist/hydrogeologist. All of the criteria for selection of where monitoring wells are required and when they need to be sampled are based upon the geology and the hydrogeology of the site. Too many of our projects have followed a pattern of the geologists and hydrogeologists getting off the project once the investigation is finished. Monitoring is part of the project, and as can be seen in the calculation above, a very significant part of the project.

Monitoring programs should be based upon the rate of the groundwater movement, the physical and geochemical properties of the aquifer, and the biogeochemistry of the contaminants in the plume. I have severe problems with monitoring plans that require regular, short-time interval sampling when conditions at a site do not warrant such costly procedures. Slow-moving groundwater, high organic content soil, organic compounds with high retardation, highly degradable organic compounds, and many other conditions all tend to make plumes move very slowly or not at all. When any of these conditions are present at a site, there is very little need for monthly or quarterly sampling for the life of the remediation. Depending on the magnitude of the conditions found at a site, yearly or every other year sampling should be more than adequate.

The original investigation should provide enough data to determine the amount of sampling needed to ensure that the remediation is performing as designed. In fact, one of the goals of the original investigation should be the collection of the data needed to design the monitoring plan. Probably the most important aspect is to determine if the plume is moving. If the plume is not moving and the remediation is going to take 20 years, then short-term monitoring is not going to provide any benefit to the operation.

While the Project Manager section of this article will provide several suggestions on specific methods to develop a program to reduce monitoring, only the person with training in geology/hydrogeology can provide the needed technical basis for the implementation of these ideas. Monitoring programs have to be based on scientific methods, not just costs. It is easy for me to say that we should reduce monitoring; the hard job is left to the geologists/hydrogeologists.

RESEARCH: Provide a scientific basis for reduced monitoring.

One of the large holes in our knowledge in the remediation field is the scientific application of monitoring well data. Of course this lack of knowledge goes beyond the simple criteria of less monitoring that we have been discussing here. There are many other areas that go into proper monitoring programs including: well construction, screen placement, well development, large flow flushing of wells before sampling, low flow sampling techniques, and many others. However, we see a significant amount of research and papers in all of these areas. People are working on improving the way that we are gathering data.

The only area in which we do not see any significant research is the timing area. Maybe I am reading the wrong journals or going to the wrong conferences (or playing too much golf), but I have not seen many papers on monitoring programs. This is surprising in that we have an enormous database on which to do research. Think about all of the public domain sites (Superfund) that have had many years of monitoring data collected, recorded, and submitted to the EPA. These data are all available for someone to go in and compare the results from multiple monitoring plans. Quarterly sampling could be compared to yearly sampling, and the variation in results could be statistically evaluated. Different types of monitoring timing could all be compared. There is probably enough data to perform these studies on multiple geological conditions. This would be the start of a firm scientific footing for monitoring programs.

The research assignment is the easiest one. I want several universities and/ or technical associations to assign, cajole, force, beg (you get the idea) some students or other underpaid personnel to perform some statistical analysis of existing data and determine the significance of quarterly sampling. These studies then need to be put in writing, and published or presented. This set of papers would then be available for the regulators to use as a basis to review monitoring plans submitted to them.

REGULATORS: Stop asking, "I wonder what is happening over there" and switch to, "What am I going to do with the data?"

The local, state and federal regulators probably have the toughest assign-ment of all of the groups. The main objectives of this group are usually: (1) protection of human health and safety, and (2) return natural resources (soil and aquifer) to their original conditions, or, at least, to usable conditions. Of course, there are 3,000 to 5,000 pages of regulations that state the same thing in a little more detail.

These regulations do state that the regulator should consider costs when specifying a remediation design. But, usually, it is not explicitly stated that costs should be part of the monitoring selection process. I have always con-sidered that monitoring was part of the remediation design, and so cost-effec-tiveness should be one of the criteria. Others see monitoring as the next stage of the process and completely separate from the remediation. Whatever the background, the regulator has been left on its own when deciding where costs figure into the monitoring selection process.

Even if there is a desire to reduce costs, the regulator has very little scien-tific information to base decisions upon. As we discuss in the Research sec-tion above, there is very little published work on optimum design of monitoring plans. Most regulators find themselves in a position of being presented with monitoring plans with very little scientific backup for the plan. They have to decide if the proposed monitoring will be sufficient to protect the public and ensure that the sites are really making progress.

I have a few suggestions for the regulators. First, I suggest that we stop looking at a site and asking, "I wonder what is happening over there?" I have seen this approach taken with most investigations by geologists and regula-tors. However, most spills are old spills, and most plumes are not moving even before we start to implement our remediation plan. Also, most ground-water moves very slowly. Accepting all of this, we should switch our ques-tion to, "What I am going to do with the data?" Immediate confirmation that the remediation is having its desired effect is understandable. Once that has been established, there is very little need for numerous sampling events. What will you do with the data? Based upon the rate of groundwater movement and/or plume movement, the progress of the plume can be watched on a yearly basis, if not every other or third year.

Another way to think about monitoring is to use event monitoring more than period monitoring. Use annual monitoring as a regular method and then use unusual events to precipitate monitoring at the site. The events can be anything from an annual sample showing high levels of contaminants, or the remediation system having mechanical problems for an extended period of time, or a hundred year rainstorm. Any of these events could change the conditions at the site. An additional monitoring round would provide the necessary confirmation that everything is all right or the data that would be required in order to react to the new conditions.

PROJECT MANAGERS: Use some of the following suggestions to help reduce monitoring events.

There are a variety of ways to reduce monitoring events, but we only have room to touch on a few here. Two key factors in successfully reducing all monitoring programs are: (1) continually evaluating the Long Term Monitoring (LTM) program to ensure that it is consistent with program objectives and decision rules, and (2) getting regulatory acceptance to the LTM optimization prior to its implementation. The best use of this factor would be to design the monitoring reduction into the original remediation design. The following specific suggestions are based upon the author's experience with remediation designs and regulatory agencies. All had input from many individuals from numerous remediation teams.

Suggestion #1

WIND DOWN THE SAMPLING PROGRAM. In general, the idea behind this suggestion is to plan to reduce the number of sampling events as the project progresses. Quarterly sampling can be used at the beginning to ensure that the remediation is performing as designed. After three or four quarters of sampling, the monitoring could be performed only twice per year. Finally, after one or two years of biannual sampling, the monitoring could be switched to once per year or every other year.

Even if we are seeing variation in the quarterly samples, the wind-down program can be implemented. Let's say the site has been sampled quarterly for the past two years and concentrations rise somewhat during the June sampling event both years. During the other six events, concentrations remain relatively constant or drop. The project manager should consider reducing sampling frequency to once or twice per year, making sure to include the June event. Or in other words, identify the critical season(s) when contaminants of concern (CoC) have the highest potential for an adverse impact.

While we are on the subject of seasons, are seasons the basis for quarterly sampling? If that's the case, then why do we sample quarterly in climates that have little if any change throughout the year? Talk among yourselves.

Suggestion #2

REDUCE THE NUMBER OF WELLS being sampled. For sites that have undergone complete investigations and/or for sites that have established that the plume is stable, a review of the number of monitoring wells should be performed. There are several reasons to eliminate a particular well from a sampling event or from all future sampling events. Wells located away from potential pathways leading to a receptor, wells located in close proximity to each other and sampled within the same interval, wells screened over multiple aquifers, and poorly constructed or damaged wells should all be considered for elimination. Any time a well provides unnecessary, redundant, or unreliable information, don't sample it, and consider eliminating it.

In addition to these technical reasons, there is one more basic reason. Most sites have more than enough monitoring wells. Simply, there is no reason to sample 20 wells every quarter just because the wells exist and they were sampled last time. What are you going to do with the data? Evaluate each well every time a sampling event is planned and eliminate some of them.

Suggestion #3

REDUCE THE ANALYSIS PROGRAM. ONE other way to reduce the cost of a sampling program is to reduce the costs from analysis. First, be sure that CoC have been properly identified. The sampling program does not have to always include nonessential compounds in the analysis program. Secondly, the number of analytes deemed "necessary" at the beginning of the monitoring program may change over time. When a CoC repeatedly reaches an agreed-upon threshold level or nondetect, stop analyzing for it. Or, if CoC is not seen along pathways that lead to a receptor, consider omitting it.

This is especially useful when one of the compounds is not part of the main family of contaminants. For example, if the main contaminants are all VOCs, and one compound is an SVOC. If and when the SVOC goes below detection limits, the entire SVOC analysis can be eliminated from the sampling program. This is especially true when dealing with metals. The metals do not usually travel as far as organic compounds in the aquifer and they decrease in concentration quickly. Once it is determined that the metals are not one of the contaminants in that part of the plume, the metal analysis can be eliminated from the sampling program.

Suggestion # 4

ABANDON WELLS THAT ARE NO longer required in the LTM program. The easy decision is when wells are old or in poor physical condition. These wells should be eliminated. The harder decisions come when we evaluate relatively new wells or wells that are in good shape. If the data that they can produce are no longer needed, then we should evaluate the elimination of these wells also. Leaving old wells exposes the owner to liabilities from potential dumping and other unexpected transport from the surface directly to the aquifer. Unneeded wells should be abandoned as soon as possible. Eliminating wells is also one sure way to reduce monitoring. One thing we lose sight of is there is an appropriate time to take your toys and go home. Think endpoints.

Suggestion # 5

USE AN INDICATOR PARAMETER INSTEAD of the entire suite of organic analysis. It is less expensive to analyze for one compound than to run an analysis for all organic compounds. This is especially true if the analysis includes runs for VOC, SVOC, high molecular weight organic, and inorganics. If the plume is stable, then a single compound may be a good way to watch the progress of the remediation. Even more money can be saved if a field parameter can be substituted for a full sampling and analysis program. An example would be substituting oxygen measurements in the vadose zone for organic analysis. If the technology for remediation was bioventing, instead of measuring the organic concentration, increase or decrease of oxygen could be used to follow the progress of the remediation. Oxygen is a field measurement and is significantly less expensive than any organic sampling and analysis. Other indicator measurements can be used depending on the specific situation.

Suggestion # 6

ALTERNATE THE WELLS TO BE sampled. Even if your monitoring program calls for quarterly sampling, this does not mean that every monitoring well has to be sampled every quarter. By alternating the wells that have to be sampled each sampling period, the total number of wells tested per year can be reduced while still maintaining the entire monitoring well network. If the wells at a site were divided into four groups, and each group of wells was sampled each quarter, then the resulting program would be the same as switching to yearly sampling. Of course, this program plan can used as part of a wind-down program as discussed in Suggestion 1.

CONCLUSIONS

AS YOU CAN SEE, MONITORING is a significant part of the remediation process. This is true technically and economically. It is going to take all of us to improve the technical, economic, and regulatory application of monitoring. You have all been given your assignments.

The suggestions should get the project managers going. I know, they aren't applicable for each and every situation but we need to use practical thinking if we are to start to reduce costs. The regulatory and research portions of the program will also have to begin as soon as possible.

Finally, we need to talk about the compensation program. Since the magazine will not collect money for me, you are going to have to send the checks directly. I think that 0.1% of savings is a reasonable figure. I know that I can trust all of you with your own calculations. Of course, I consider a simple "thank you" or "good article" worth as much as $1,000. That means if the program suggested above is 25% effective, I expect 5,000 "attaboys." I promise to share all nice comments with Lynne.

PART II

PUMP AND TREAT
REMEDIATION

8

Using Laboratory Data to Design a Full-Scale Treatment System

Evan K. Nyer

INTRODUCTION

I WAS RECENTLY ASKED TO review a Request For Quotation (RFQ) for a laboratory treatability study. After reading through the RFQ, I decided that it would be a great basis for a general article on the proper use of laboratory data and studies.

Most consultants envision a laboratory study as a scale model of the full-scale treatment system. They feel that as long as the correct scale factor is employed, any full-scale unit can be simulated by a small laboratory unit. The data generated by the laboratory equipment can then be directly used to design a full-scale treatment system. This perception is wrong.

The laboratory cannot be used to directly simulate full-scale treatment units. Laboratory studies can only be used to develop information about physical, chemical, and biochemical reactions. Most full-scale unit designs are based upon solving practical problems that occur when one is trying to apply these reactions in the field. A good full-scale design is derived from a combination of the desired reaction and the solution to the practical problems.

In addition, laboratory reactions do not always simulate field reactions. For this reason, all reactions cannot be performed in the laboratory. Some reactions are very accurate. Metal precipitation would be one example of a reaction that can be accurately simulated in the laboratory. Other reactions will not perform the same in the laboratory as in the field. Air stripping is a good example of a technology that performs differently in the laboratory and the field.

The best way to explain both of these phenomena is to review the RFQ. The laboratory can be a powerful tool in the design of full-scale treatment systems. However, we first have to understand what the data actually means, and how it can (or cannot) be applied to a full-scale unit.

FULL-SCALE TREATMENT SYSTEM

THE PURPOSE OF THE LABORATORY study in the RFQ was to simulate a full-scale treatment system that was planned for a site. It was desired to generate some operational data before investing in the full-scale treatment system.

Table 8.1 summarizes the organic compounds detected in the water at the site. The RFQ also expected to have about 5 to 10 mg/L of iron in the water. Figure 8.1 summarizes the treatment system that the RFQ wanted to employ to treat the water. As can be seen in Figure 8.1, the unit operations are: (1) iron oxidation by aeration, (2) suspended solids removal by sand filtration, (3) volatile organic removal by packed tower air stripping, and (4) general organic removal by activated carbon adsorption.

For each unit operation a laboratory test was specified. Let us review each test and see how the data was supposed to be used.

IRON OXIDATION

THE FULL-SCALE OXIDATION SYSTEM consisted of a reaction tank with a one-hour residence time. A blower forced air through an air distribution system. The air provided the oxygen for oxidation and the mixing for the tank. Small bubble air diffusers were used to maximize oxygen transfer.

The laboratory test was supposed to be set up to simulate the full-scale system. A continuous flow tank with a one-hour residence time was to be set up. Air flow and dissolved oxygen content was to be monitored to ensure that sufficient mixing and oxygen transfer would occur in the full-scale treatment system. Dissolved iron concentration at selected air flow rates was the main variable to be monitored.

Several problems exist with this testing procedure. First, air flow for mixing and oxygen transfer cannot be combined and scaled to a full-scale system. The scale factor for oxygen transfer is mainly affected by the depth that the air is released in the tank. In general, this parameter has a linear relationship

TABLE 8.1

Contaminant	Concentration, μg/L
Carbon tetrachloride	50
1,1-Dichloroethane	65
1,2-Dichloroethane	35
Methylene chloride	160
1,1,1-Trichloroethane	95
1,1,2-Trichloroethane	70
Trichloroethylene	230

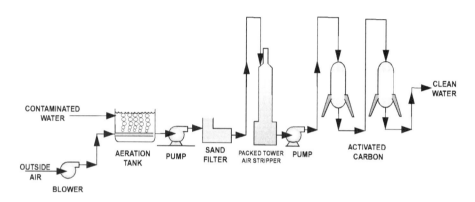

FIGURE 8.1. Water treatment system.

when extrapolating to a full-scale system. Mixing efficiency is related to the total power input into the tank, and has an exponential relationship with the full-scale system. Neither of these critical design factors should be based upon laboratory data. Both should be based upon experience with full-scale designs. The laboratory test should only be used to test the effect of oxygen concentration on the iron oxidation rate. How you get that oxygen concentration is very different for laboratory units and full-scale units.

There is one other small problem with this test. Large water samples cannot be returned from the field without introducing oxygen. While we may pride ourselves in being able to take samples for volatile organic analysis without air space in the sample containers, we are not going to be able to perform that bit of magic with a 55 gallon drum. Even if there is no headspace, there still will be small amounts of oxygen transferred into the sample. This oxygen will react with the iron in the sample. All of the iron will be oxidized before the start of the laboratory tests.

The only way to test the rate of iron oxidation on a groundwater sample is to perform the test in the field. The test must be run as soon as the water is taken from the well. Even in the field, great care should be taken to avoid the introduction of oxygen into the water sample when removed from the well. As a side note, I would suggest that pH be measured during the test. The pH has a dramatic effect on the rate of oxidation. The lower the pH, the slower the rate of reaction.

FILTRATION

NOW THAT WE HAVE TAKEN our soluble iron and turned it into insoluble iron (suspended solids), we only have to remove the solids in order to remove the iron from the water. The present system planned to use a sand filter for the solids/liquid separation step of the treatment system and the laboratory study was to simulate a sand filter.

The prescribed method in the RFQ was to set up a column with sand and to test the effect of various flow rates on the removal of suspended solids and iron. There are detailed specifications to ensure that the tests were accurate. For example, the column had to be wide enough to ensure that there were minimal wall effects; i.e., 30 times the size of the sand particles.

The details were very interesting, but had very little to do with the operation of a sand filter. Sand filters do not strain suspended solids from water like filter paper does. Sand filters, and all particle filters, remove the suspended solids by solid/solid contact. The suspended solids come into contact with the sand particle and attach. The process has a closer relationship to flocculation than to straining of suspended solids.

The only time that a sand filter acts as a strainer is when the suspended solids have filled all of the void spaces between the sand particles. At this point, the pressure required to force the water through the bed is too high and the filter is usually backwashed. This removes the suspended solids and opens the pore spaces once again.

The efficiency of the sand filter is related to the stability of the suspended solids. If the filter is not removing the required percentage of the suspended solids, then the chemistry of the water is changed, not the water velocity. A flocculent is usually added to the influent of the sand filter to remove more suspended solids. These results cannot be simulated in a laboratory scale unit. And even if these tests could be performed, they would not simulate the actual operation of the sand filter.

The main problem with the operation of the actual sand filter is cleaning it. Once the suspended solids are separated by the filter, can they be removed from the sand particles again? Sand filters usually fail because they cannot be cleaned, not because they are not removing suspended solids.

Even though the laboratory scale unit looks like the full-scale unit, it cannot perform the same as the full-scale unit. Even with scale factors, the laboratory unit cannot simulate or predict the operation of a full-scale sand filter.

If it is necessary to remove a certain size or type of suspended solid in a treatment system, then a sand filter is the wrong unit operation. In those circumstances, the designer should rely upon a different type of filter; i.e., cartridge filter, ultrafilter, etc.

PACKED TOWER AIR STRIPPER

A FULL-SCALE PACKED TOWER air stripper cannot be simulated in the laboratory. The key to this type of air stripper is the performance of the packing. The normal packing used in a tower is between 2 to 3 in. in diameter. Unfortunately, we cannot use smaller media and a scale factor, and we cannot afford to process sufficient water for media of this size in a laboratory setting.

The contractor must have realized this and did not provide specifications on the air stripper test. The RFQ simply stated that the air stripper should be simulated in the laboratory.

The only way to test an air stripper is to perform a pilot study in the field. The tower should be a minimum of 8 to 12 in. in diameter, and 7 to 10 ft of packing depth. The testing should include running the air stripper from 10 to 30 gpm/ft^2. For an 8-in. pilot tower, a range of 3.5 to 10.5 gpm would be required to fully test the packing. In addition, several runs of 30 to 60 min duration would be required to develop the data necessary for the full-scale design. As can be seen, a significant amount of water is required to test this technology.

There is one other method to acquire the necessary data. Most companies that sell air strippers have developed computer models that simulate the air stripper's performance. In addition, commercial computer models are also available for this purpose. As long as an accurate Henry's law constant is available, then these models can provide the required data on air stripper performance in a particular situation. If the input data are not available, then a pilot plant is the only way to develop that performance data.

ACTIVATED CARBON

THE SPECIFICATIONS FOR THE LABORATORY-activated carbon unit were 30 gpm/ft^2 and 30 min residence time. When I first received the RFQ, I thought that this was a typographical error. A 120-ft tall column would be necessary to meet these specifications. You may perform your own calculations to confirm this.

After calling the author of the specifications, I found out that they were simply following the specifications that were listed with the full-scale carbon unit. The problem is that the 30 gpm/ft^2 refers to the design limitations of the tank and piping, not the carbon. The 30 min residence time refers directly to the carbon itself. The RFQ combined these two into one specification for the carbon. After some detailed discussion, it was decided to use 3 gpm/ft^2 as the basis of the carbon system. This would be a normal flow rate for a field carbon unit.

This left one problem with the laboratory test. We calculated the amount of contaminated water necessary to break through the activated carbon in the laboratory unit. This can be done by using the isotherm data provided in the Fall, 1991 issue of this column in *Ground Water Monitoring Review*. We cannot measure the performance of the carbon column until the compounds start to come through the unit and can be measured. We estimated that 5,000 to 6,000 gallons of contaminated water would be required for the test.

I would like to know if anyone has ever approached a government agency and told them that they were going to remove 5,000 gal. of untreated water from a Superfund site and send it to a laboratory. Of course, this is too much water to bring into a laboratory. The other thing to remember is that all of this water would have to be pretreated with the rest of the treatment system in order to perform an accurate simulation, and the water would have to be disposed of in an appropriate manner.

There are three choices when performing tests on activated carbon. First, isotherms can be performed in the laboratory. These data are easy to obtain, but are not always accurate when compared to activated carbon in a column. The second choice is to perform a field pilot test with a small column of activated carbon. The data are very accurate, but the test can get complicated if pretreatment is required. The third choice is to send a sample to Calgon Carbon. They have developed a laboratory test using powdered activated carbon in a column that simulates full-scale units.

There is one other method to get the required data. Most of the companies that sell activated carbon have developed computer models that simulate the performance of carbon in a column. These computer models usually work for individual compounds or mixtures of organic compounds. These models can be used to design most of the carbon applications for the field. The carbon companies will also know when the models should not be used and field data is required. The models are probably more accurate than any laboratory test. The laboratory should be used to test for compounds and conditions that will interrupt the performance of the carbon. I have seen more carbon columns ruined by suspended solids than by the organic content of the water.

GENERAL SPECIFICATIONS

THE LAST SPECIFICATION OF THE RFQ was the requirement that all of the sample had to be stored at 4°C while it was waiting for laboratory processing. In general, this is a good idea, but we should evaluate whether the practical problems this may cause do not outweigh the technical results. The main benefit that the low temperature provides is to ensure that there is no biological activity while the sample is being stored. Because the primary organic compounds in this study are not degradable, the low temperature storage will not gain that much accuracy. On the other hand, storing 5,000 to 6,000 gallons of water at 4°C will be difficult.

SUMMARY

THE LABORATORY PROVIDES A POWERFUL tool that can be used to derive costs and answer technical questions concerning water treatment design. The problems described in this article come from the misconception that all full-scale processes can be simulated by laboratory-scale treatment units. This is simply not true. We can always generate data from a laboratory unit, but we must make sure that the data are applicable.

When starting to design a treatment system, there is always a requirement for data to be used as a design basis for the unit operations. We need to define exactly what data are required, and then decide what is the best way to generate that data. We need to use a combination of laboratory studies, pilot plants, literature, and experience in order to develop an accurate and productive design.

9

Treatment System Operation and Maintenance: Critical Factors in an Economic Analysis

Evan K. Nyer and Gregory Rorech

INTRODUCTION

WE WERE RECENTLY ASKED TO install a treatment system that had been designed by another consultant. After review of the design, we suggested that we be allowed to make changes before construction commenced. The objective of the system was to remove benzene, toluene, ethylbenzene, and xylene (BTEX) from groundwater. The projected cost for the original system over the 20-year life of the project was $19,124,000. After we had completed our changes, the projected cost of the remediation was $7,563,000.

The surprising part of the savings was that we reduced the capital costs by only $1,686,000. The rest of the savings for the project were the result of changes to the operation and maintenance of the original treatment system. The design changes that we suggested would result in a savings of $9,875,000 in operation and maintenance costs over the life of the project.

It has been said many times in this column that anyone can design a $20,000,000 treatment system. All you have to do is select every treatment technology that will remove the compounds found in the groundwater and place them in series.

One of the important lessons to be learned from this project is that neither treatment system was "wrong." Both systems would have removed the BTEX and met all regulatory criteria. Both would have remediated the site. The only difference is that one system design cost $11,561,000 less than the other! To see why, we'll review the site and the systems.

THE CONTAMINATED SITE

THE SITE WAS A PETROLEUM bulk storage terminal located in northern Florida. A residential community is directly north of the site with a river and light industrial complexes found on the other borders. Past incidental releases of product from interconnecting pipelines and tank bottom water releases had impacted the surficial aquifer. Nearby residential irrigation wells were impacted and later abandoned. No additional releases have occurred during the past seven years.

Groundwater and soil investigations have been underway since 1986. The site is underlain by an unconfined sand aquifer and a semiconfined limestone aquifer, separated by a clay-rich layer that functions as an aquitard.

THE ORIGINAL PROPOSED TREATMENT SYSTEM

THE ORIGINAL PROPOSED TREATMENT SYSTEM is shown in Figure 9.1. The system consisted of a flow equalization tank; two air strippers in series; vapor phase carbon for the off-gas from the air strippers; two sand filters, operated in parallel; solids dewatering equipment; and two activated carbon units in series. The capital costs are presented in Table 9.1, including each unit, the recovery wells, piping, site preparation, equipment installation, electrical power, permits, engineering, and project management. The total capital cost for the original system was $4,357,000 including a 20% contingency.

The operating costs for the original system are summarized in Table 9.2. These costs include power, air stripper cleaning (iron in the groundwater), sampling and analysis for the treatment system and the monitoring wells, maintenance for the equipment, liquid phase and air phase carbon replacement, and operator expenses. The yearly operating costs for the original system totaled $1,129,000, including a 20% contingency.

This system would have collected all of the contaminated groundwater, removed the contaminants, and discharged clean water. The site would have been cleaned in 20 years. All government requirements would have been met and human health and safety would have been protected over the entire life of the project.

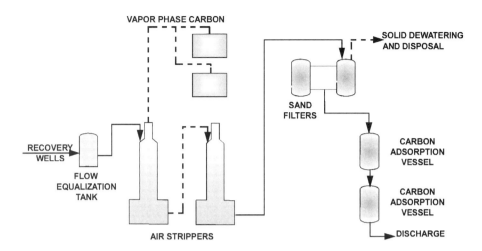

FIGURE 9.1. Original proposed treatment system.

TABLE 9-1. ORIGINAL PROPOSED TREATMENT SYSTEM

Capital Equipment	
Recovery wells and piping	$446,000
2 Air strippers	280,000
Air treatment system	190,000
Filters	220,000
Carbon adsorption	140,000
Vapor extraction system	190,000
Site prep, installation	1,020,000
Electrical, power	411,000
Project management, engineering permits	734,000
Contingency, 20%	726,000
Total	$4,357,000

NEW TREATMENT SYSTEM

THE FIRST STEP THAT WE took after receiving the project was to review the strategy. Many of the design decisions that we make for a treatment system are based on assumptions made during the investigation of the site.

After the review, we decided that there was insufficient information on which to base the placement of the recovery wells and monitoring wells. Originally, a 300-gpm treatment system was specified, based on one short-term

TABLE 9-2. ORIGINAL PROPOSED TREATMENT SYSTEM

Operations and Maintenance Per Year	
Power	$121,000
Air stripper cleaning	18,000
Sample and analysis–treatment system	49,000
Sample and analysis–wells	522,000
Maintenance	36,000
Liquid phase carbon	55,000
Air phase carbon	34,000
Operator	106,000
Contingency, 20%	188,000
Total	**$1,129,000**

pumping test and slug test data. Also, the treatment system was complex and designed conservatively. Based on the potential cost savings, we asked that we be permitted to conduct additional activities to refine the design.

They included:

- perform two long-duration aquifer performance tests
- obtain updated water-quality data to develop modifications to the original design
- perform 3-D modeling with the updated hydraulic characteristic data determined from the aquifer performance tests.
- develop modifications to the design of the recovery and treatment system.

The 3-D model showed us that the plume could be captured with fewer recovery wells, and that the remediation could be monitored with fewer monitoring wells. The model also allowed us to predict the BTEX concentrations that we could expect at the treatment system over the life of the project.

Based on this refined information, we were able to reduce the number of recovery wells from 14 to 6, and the flow to the treatment system from 300 gpm to 145 gpm. We were also able to show that the BTEX concentration would be low enough once all of the wells were pumping to extend the stack of the air stripper to meet air permit requirements without vapor-phase treatment.

Figure 9.2 summarizes the unit operation for the revised treatment system. The system includes an aeration tank (to precipitate the iron); a continuous sand filter (to remove the suspended solids and protect the air stripper from fouling); an air stripper; and a recharge system. The iron pretreatment system is included in this cost analysis, but will not be installed until after the air

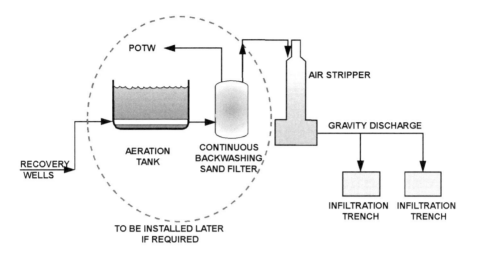

FIGURE 9.2. New treatment system.

stripper is operated for a period of time. We will analyze whether it is less expensive to clean the air stripper periodically or if we should pretreat to remove the iron during this operation. The groundwater data and the model were not able to give us sufficient information to make this decision before construction and operation. The treatment system will be built so that these unit operations can be added easily at a later time. The capital costs for the new treatment system are summarized in Table 9.3. The same costs that were included in the original treatment system are included in this system with the addition of the costs for modeling. The total capital cost is $2,671,000.

The operating costs are summarized in Table 9.4. The costs are the same as those for the original treatment system with two categories eliminated. First, the new treatment system will not require activated carbon for the air stream or the water stream, so carbon replacement costs are not included in the new system operation. Second, the iron, in the form of suspended solids, will be sent to the local Publicly Owned Treatment Works (POTW) for disposal; there will be no BTEX coming from the continuous backwashing sand filter effluent. The yearly operating costs for the new treatment system are $374,000, including a 20% contingency.

COMPARING THE TWO TREATMENT SYSTEM DESIGNS

TOTAL COSTS ASSOCIATED WITH BOTH treatment system designs are summarized in Table 9.5. In order to perform a direct comparison, costs, as shown in Table 9.5, are based on present value. To calculate the values for the operat-

TABLE 9-3. NEW TREATMENT SYSTEM

Capital Equipment	
Recovery Wells and Piping	$258,000
Air stripper	91,000
Aeration basin	70,000
Continuous sand filter	140,000
Site prep, installation	357,000
Electrical, power	110,000
Project management, engineering, permits, modeling	1,200,000
Contingency, 20%	445,000
Total	**$2,671,000**

TABLE 9-4. NEW TREATMENT SYSTEM

Operations and Maintenance Per Year	
Power	$42,000
Air stripper cleaning	15,000
Sample and analysis - treatment system	24,000
Sample and analysis - wells	83,000
Maintenance	24,000
Sand filter discharge to POTW	33,000
Operator	91,000
Contingency, 20%	62,000
Total	**$374,000**

TABLE 9-5. TOTAL COST FOR TREATMENT SYSTEMS

	Capital	Operating[a]	Total
Original proposed treatment system	$4,357,000	$14,767,000	$19,124,000
New treatment system	$2,671,000	$4,892,000	$7,563,000

[a] Present value: 20 years at 5%.

ing costs in Table 9.5, we took the yearly operating costs from Table 9.2 and 9.4, and calculated the present value based upon a 20-year life of the project and 5% value of money. The total savings using the new design are $11,561,000.

The main savings came from the operational changes resulting from the new design. More than 85% of the savings came from changes to operation and maintenance. The main decreases in operation and maintenance came from three changes. First, the number of monitoring wells and their sampling frequency were decreased significantly. Second, the carbon replacement was eliminated. Third, the total system was smaller and simpler, allowing for a general reduction in power, man power, and maintenance.

The only increases came from the discharge to the POTW and the modeling efforts. Both items resulted in significant savings in capital and operating costs with the new system.

SUMMARY

IT IS EASY TO WASTE $20,000,000. This case history shows that nothing is gained by simply throwing money at a problem. All components must be considered when designing a remediation system. It is important to have a full understanding of the groundwater and the nature of the contamination. Involve someone with experience when designing the treatment system, and then extra equipment will not be needed "just to make sure." Finally, analyze the capital and operating costs of the treatment system over the entire life of the project. The operating and maintenance costs can be the critical factor in the economic analysis of a treatment system.

10

TRICHLOROETHYLENE TREATMENT AND REMEDIATION

Evan K. Nyer and Bridget Morello

THIS SECTION USUALLY REVIEWS A specific treatment technology. For a change of perspective, we would like to change the format and concentrate on a single compound. However, in trying to write this article, we found that there was a small language problem: many people use the terms remediation and treatment interchangeably. We consider these terms to represent very different concepts. Treatment is the application of a technology to a specific medium for the separation and/or destruction of a compound. We treat an air, water, or soil medium. Remediation is the cleaning of an environment. We remediate a wetlands, an industrial site, or a Superfund site. For example, we treat groundwater, but we remediate an aquifer. When considering chlorinated hydrocarbons, we have come a long way in treatment methods. However, we still have a long way to go when we discuss remediation of a site contaminated with chlorinated hydrocarbons. Let us use trichloroethylene (TCE) as an example to illustrate this point.

During the past 10 years we have gathered vast experience in the treatment and remediation of TCE contamination in groundwater and soils. We now have complete understanding of the technical requirements for treating this compound in air, water, and soil. However, we do not possess easy solutions to all of the technical limitations associated with remediating TCE in various environments. Current TCE remediation methods rely on water or air as a carrier to remove TCE from the aquifer, vadose zone, and soil. Although the treatment of TCE in water, soil, and air is relatively straightforward, remediation is limited because residual concentrations of TCE in the groundwater or soils often reach asymptotic levels that are greater than the required TCE cleanup standards (usually health-based standards for exposure to water and soil).

To date, the main treatment alternatives for removing TCE from groundwater have been: (1) air stripping, (2) granular activated carbon (GAC) adsorption, and (3) ultraviolet (UV) oxidation. A new method is the use of biological reactions to treat TCE contaminated soils and groundwater aboveground (ex-situ). Let us review each of these treatment methods.

Air stripping is the process of chemical transfer from an aqueous phase to a gaseous phase. The controlling factor in the removal of volatile organic compounds (VOCs), like TCE, is the rate of transfer from the groundwater to the air until equilibrium is established. Typical air strippers employ a countercurrent configuration which removes VOCs by subjecting a stream of water flowing downward through packing to a mass of air flowing upward through the packing, thus allowing intimate contact of water and air. Figure 10.1 illustrates a typical air-stripping system. The design of an air-stripping tower is guided by the level of each constituent in the water subject to treatment, its Henry's law constant [544 atmospheres at 68°F for TCE (Nyer, 1992)], and the anticipated air-to-water ratio for operation. Limitations to air stripping include: (1) fouling, scaling, and biological growth on tower internals; (2) water channeling through the packing due to unequal water distribution; and (3) flooding which occurs when the downward flow of water is restricted or impeded by the air flow. New air-stripper designs that help solve some of these limitations include diffused aerators, tray aerators, rotary strippers, and fluidized beds. Additionally, due to severe limitations on air emissions in this country, most air-stripping applications require further treatment of the stripper off-gas. The two main methods of air-stripper off-gas treatment are carbon adsorption and catalytic oxidation. In some cases, where very low concentrations of TCE are present, no air-stripper off-gas treatment is required.

One example of air-stripping treatment is the system which Geraghty & Miller, Inc. designed for removal of TCE at a manufacturing site located in Stuart, Florida. The operations at this site utilized solvents and paints, and included storage of aviation gasoline and jet fuel. The groundwater in the shallow aquifer at this site had become affected with volatile organic compounds, including large amounts of TCE. As a result of the solvent releases to groundwater, several City of Stuart municipal water supply wells were impacted. The treatment system was sized for a 1,000 gallon per minute (gpm) flow rate and designed to provide complete air-stripping of VOCs (primarily TCE) in order to use the groundwater for the municipal water supply. The

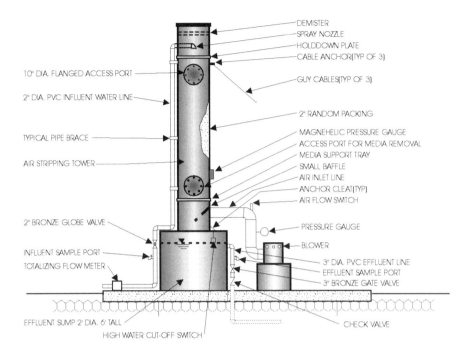

DEMISTER
SPRAY NOZZLE
HOLDDOWN PLATE
CABLE ANCHOR(TYP OF 3)

10" DIA. FLANGED ACCESS PORT

2" DIA. PVC INFLUENT WATER LINE

GUY CABLES(TYP OF 3)

2" RANDOM PACKING

TYPICAL PIPE BRACE

MAGNEHELIC PRESSURE GAUGE
ACCESS PORT FOR MEDIA REMOVAL
MEDIA SUPPORT TRAY
SMALL BAFFLE
AIR INLET LINE
ANCHOR CLEAT(TYP)
AIR FLOW SWITCH

AIR STRIPPING TOWER

2" BRONZE GLOBE VALVE

PRESSURE GAUGE

INFLUENT SAMPLE PORT
TOTALIZING FLOW METER

BLOWER

3" DIA. PVC EFFLUENT LINE
EFFLUENT SAMPLE PORT
3" BRONZE GATE VALVE

EFFLUENT SUMP 2' DIA. 6' TALL
HIGH WATER CUT-OFF SWITCH

CHECK VALVE

FIGURE 10.1. Air stripper detail.

system consisted of two air strippers each 7.5 feet diameter and 35.5 feet tall, with 22 feet of random packing. The influent TCE concentration was 2400 parts per billion (ppb). In order to achieve the required cleanup level of 3 ppb, the air-stripping system was designed for a TCE removal efficiency of 99.87%. Since startup, the air-stripping system at this site has effectively reduced TCE to nondetect levels from the contaminated groundwater. Since maximum acceptable air quality concentrations were not exceeded, no off-gas treatment was required at the site.

Another commonly used technology for removal of VOCs, like TCE, from groundwater is GAC adsorption. Adsorption occurs when liquid molecules are attracted to and held at the surface of a solid. Activated carbon is an excellent adsorbent because it has a large internal surface area available for adsorption. Large openings, or macropores, provide an entrance to the internal surface area of a carbon particle where adsorption takes place. Liquid-phase adsorption isotherm tests and dynamic column studies are often used to assess the feasibility of activated carbon adsorption for a specific application. Isotherm tests demonstrate the degree to which a particular dissolved compound is absorbed onto activated carbon; the adsorption capacity for TCE is

18.2 milligrams TCE per gram carbon at 500 ppb (Nyer, 1992). Dynamic column studies are the only method of determining the optimum contact time and mass transfer zone which depend on the rate at which constituents are adsorbed on the carbon. Limitations to GAC adsorption are usually economic, specifically in relation to carbon consumption rates and carbon regeneration/disposal. GAC adsorption is not typically employed as a stand-alone treatment system, but rather as a polishing process in conjunction with other treatment units (e.g., air strippers). An example of GAC adsorption used in conjunction with air-stripping for remediation of contaminated groundwater is the Seymour Superfund site in Seymour, Indiana. The groundwater at the Seymour site was impacted by organic compounds leaking or spilled from the approximately 50,000 fifty-five gallon drums and 100 large tanks containing chemicals. The initial groundwater treatment system at the Seymour site consisted of an air stripper, multimedia filter, and GAC adsorbers; and effectively treated organic compounds, including TCE, to levels acceptable for discharge to the local POTW.

UV oxidation typically involves combining ultraviolet radiation and oxidants such as hydrogen peroxide, ozone, or both. UV radiation enhances chemical oxidation of organic compounds, like TCE, found in groundwater. Figure 10.2 illustrates a basic UV oxidation system. UV oxidation can destroy over 99% of TCE while yielding a harmless chloride ion end product. Advantages of UV oxidation over other conventional groundwater treatment technologies include: (1) contaminant destruction rather than transferral into another phase or onto some media, and (2) no associated residual generation (e.g., spent carbon, sludge). The main disadvantage of UV oxidation stems from the concentration of natural constituents (e.g., iron, calcium) in the aquifer which, due to their propensity for contributing to fouling and scaling, can adversely affect the chemical reactions occurring during treatment. Economic considerations are generally another disadvantage to UV oxidation.

Many successful UV oxidation field tests have been performed to date to evaluate system variables such as high and low intensity UV radiation, hydrogen peroxide and ozone feed concentrations, residence times, and organic loading rates. One field test performed at the Lorentz Barrel and Drum (LB&D) Superfund site resulted in up to 99% TCE removal (Topudurti and Lewis, 1991). Additionally, comprehensive UV oxidation field testing (involving various oxidation scenarios) performed at a Xerox facility in Webster, New York contaminated by chlorinated and nonchlorinated solvents exhibited 98% destruction of the organic contaminants found in the groundwater (Smith, 1991).

Based on laboratory studies, biological treatment shows enormous potential as a destruction technology for TCE found in groundwater. However, the

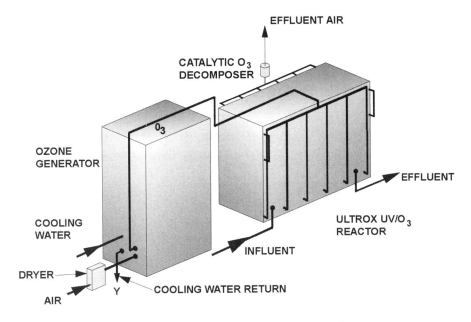

FIGURE 10.2. Typical UV/ozone process flow.

lack of documented biological field experience demonstrating achievement of cleanup standards impedes public and regulatory agency acceptance of groundwater biological treatment (OTA, 1989). Aerobic biodegradation of TCE by indigenous microorganisms has not been proved in the field.

While we have had good success cleaning TCE from water, when water is used as the carrier, we have found severe limitations in remediation. The standard pump and treat system for groundwater remediation basically uses water as a carrier to remove TCE from the aquifer. Since chlorinated hydrocarbons, especially TCE, have severe solubility limitations [the solubility of TCE is 1.1×10^3 milligrams per liter (Nyer, 1992)] only a certain amount of TCE can dissolve in the water and be removed with the water flow. Another limitation is the slow diffusion of TCE from soil particles (to which it is adsorbed) into the main water flow stream of a typical aquifer.

Traditional pump and treat systems are designed to effectively capture and remediate contaminated groundwater plume(s). Contaminant mass removal in pump and treat systems is usually significant; however, the rate of contaminant removal often declines rapidly to low levels due to a combination of three factors: (1) removing groundwater faster than the contaminants can desorb from the soil, (2) lowering water tables below the most contaminated portions of the subsurface, and (3) diluting concentrations by drawing in less contaminated groundwater from surrounding areas (Haley et al., 1989). These

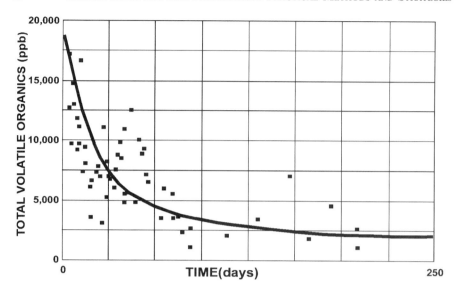

Figure 10.3. Total volatile organics versus time.

limitations, as illustrated in Figure 10.3, formulate the normal life cycle concentration pattern we have seen in pump and treat systems. The initial rapid decline in removal rates typically observed in pump and treat systems often results in the remaining groundwater having contaminant levels above cleanup criteria. Due to the limitations with pump and treat experienced over the years, the current thinking is to switch to using pump and treat for controlling plume movement and not as a major remediation method. Innovative technologies are required to treat the remaining contamination present when pump and treat systems are no longer effective in remediating the plume(s).

One alternative to using water as a carrier for the contaminants is to use air as the carrier. Soil vapor extraction (SVE) systems utilize air as the carrier to volatilize contaminants, like TCE, from unsaturated soil (vadose) zones. SVE involves moving air through the soils via vacuum to remove TCE from the unsaturated soil zone. The extracted soil vapors have to be treated when high levels of TCE are present; typically, the soil vapors are treated using carbon adsorption or catalytic oxidation. Treatment of contaminated soil below the water table is more complicated. Typical SVE systems cannot be effective at remediating these soils unless the area is sufficiently dewatered to provide unsaturated conditions capable of allowing induced air flow, and associated economic factors make this option unfeasible.

An emerging technology for treatment of contamination in saturated zones is air sparging. Sparging is accomplished by injecting air under pressure to the soils below the water table via wells constructed with perforated bottom screens. Adsorbed contaminants like TCE are volatilized by the air flow and then collected with an SVE system; or, where permissible, vapors are allowed to escape to the atmosphere. Figure 10.4 illustrates a basic air sparging system including vapor collection and treatment.

Air sparging systems operating in conjunction with SVE systems provide tremendous improvements in contaminant mass removal rates; existing data indicate that a 10% addition of air flow via a single sparge point can increase the contaminant removal 50-fold through the SVE system (Brown and Jasiulewicz, 1992). Economic savings are the main advantage in utilizing air sparging in conjunction with SVE rather than traditional treatment technologies. However, misapplication of air sparging can potentially result in disadvantages such as accelerating vapor flow to nearby receptors unless adequate venting is provided. Therefore, continued field testing and proper design evaluation are critical in determining the applicability of air sparging to various contaminated sites.

An example of successful air sparging and SVE treatment in conjunction with pump and treat (utilizing air stripping for the groundwater) is source area at a Superfund site located in the Midwest. After removal of approximately 150 gallons of nonaqueous phase liquid (NAPL), maximum background groundwater and soil TCE concentrations were approximately 49,000 micrograms per liter (μg/L) and 550,000 micrograms per kilogram (μg/kg), respectively. Operation of the groundwater pump and treat system from 1984 to 1989 reduced the groundwater TCE concentration to approximately 17,000 μg/L, and it was apparent that any further removal of TCE would be a slow process. Therefore, an SVE system was installed and operated in conjunction with the pump and treat system which resulted in a reduction in the groundwater and soil TCE concentrations to approximately 400 μg/L and 1400 μg/kg, respectively. These concentrations were still greater than the cleanup levels. An air sparging system was then installed and operated in conjunction with the SVE and pump and treat systems for approximately five months. After the air sparging system was shut down, the groundwater and soil TCE concentrations were approximately 122 μg/L and <100 μg/kg, respectively. Although a tremendous increase in VOC removal occurred upon operation of the air sparging system, TCE groundwater and soil cleanup levels (3 μg/L and 60 μg/kg, respectively) have not been achieved, and various future operating scenarios (e.g., pulsed sparging) are being evaluated.

The only alternative to using water or air as a carrier is to biologically treat the TCE in place or in situ. To date, biological treatment of soils has been evaluated primarily under laboratory conditions and in pilot-scale tests, few full-scale field demonstrations have been performed. In many cases, indig-

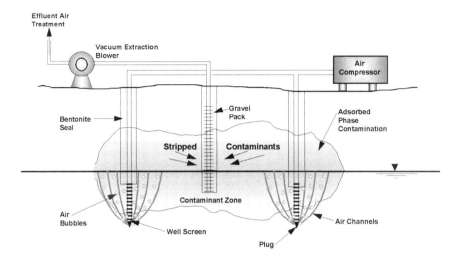

FIGURE 10.4. Typical air sparging system.

enous microorganisms possess the metabolic capability to metabolize most of the constituents present in the soils, especially when growth conditions are optimized. Generally, in-situ biological treatment has worked on petroleum hydrocarbons, but not TCE. However, TCE was successfully removed (95%) from groundwater at a California manufacturing facility by in-situ aerobic biodegradation upon introduction of a nonindigenous, laboratory-created bacterial strain called *Pseudomonas cepacia* (Mahaffey et al., 1992). The only biological treatment methods recently successfully evaluated in laboratory tests involved stimulating the natural biological population with ring compounds (i.e., benzene or toluene), or stimulating the natural bacteria with methane. These methods have also been somewhat successful in pilot-scale tests.

Inherent limitations to biological remediation of an aquifer water include the following: (1) organic contaminant loading (high concentrations can potentially poison microorganisms); (2) control of oxygen, nutrient content, pH, and temperature required to sustain biological activity; (3) natural soil and aquifer conditions (e.g., low soil permeability and low groundwater flow) which can potentially inhibit biological effectiveness; (4) regulatory agency acceptance of introducing nonindigenous microorganisms to a site (during in-situ treatment); (5) potential toxic by-products and/or adverse affects on naturally occurring degradation due to variations of existing aerobic or anaerobic conditions; and (6) testing (for discharge criteria) and subsequent disposal of metabolic by-products and cell lysis materials. For example, a study of anaerobic biodegradation at a Superfund site confirmed that TCE degradation resulted in the production of vinyl chloride as a toxic by-product (Silka and Wallen, 1988). Considering these limitations, it is imperative that further biological field testing be performed to obtain critical information regarding applicability to a wide range of sites and to establish data indicating proof of contaminant destruction.

In conclusion, TCE remediation relies on using either water or air as a carrier to convey the contamination to a treatment system. While we have been very successful at treating the carriers, the inherent process limitations often prevent reduction of original TCE concentrations to levels lower than the required cleanup standards. Innovative technologies like biological treatment and air sparging must be evaluated based on how they overcome inherent limitations associated with use of these carriers as a remediation method. Hence, treatment is easy; it is remediation that is difficult.

REFERENCES

Bernardin, F.E., Jr. UV/Peroxidation Destroys Organics in Groundwater, 83rd Annual Meeting of the Air and Waste Management Association, Pittsburgh, PA, June 24–29, 1990.

Brown, R.A. and F. Jasiulewicz. "Air Sparging: A New Model for Remediation." *Pollut. Eng.*, 24(13):52–55, 1992.

CH2M Hill. 1991. Effects of Nonaqueous Phase Liquids on a Superfund Remediation, prepared for HAZTECH International '91.

Haley, J.L., C. Roe, and J. Glass. Evaluation of the Effectiveness of Groundwater Extraction Systems. Proceedings Hazardous Materials Control Research Institute, 1989, pp. 246–250.

Loden, M.E. A Technology Assessment of Soil Vapor Extraction and Air Sparging, USEPA Project Summary, EPA/600/SR-92/173, 1992.

Mahaffey, W.R., G. Compeau, M. Nelson, and J. Kinsella. "TCE Bioremediation." *Water Environ. Technol.*, 4(2):48–51, 1992.

Nyer, E.K. and G. Skladany. Reactor Treatment Design, USEPA Bioremediation of Hazardous Waste Sites Workshop, CERI-89-11, 1989.

Nyer, E.K. *Groundwater Treatment Technology, 2nd ed.* Van Nostrand Reinhold, New York, 1992.

Silka, L.R. and D. Wallen. Observed Rates of Biotransformation of Chlorinated Aliphatics in Groundwater, prepared for Superfund '88 proceedings, 1988.

Smith, L.P. "Chemical Oxidation for Groundwater Remediation." *Water Environ. Technol.*, 3(11):34, 1991.

Topudurti, K. and N. Lewis. "New Technologies Tested for Groundwater Cleanup at Superfund Sites." *Water Environ. Technol.*, 3(11):24, 1991.

U.S. Congress Office of Technology Assessment (OTA). Coming Clean: Superfund Problems Can Be Solved, Chapter 3, 1989.

USEPA Technology Innovation Office. In-Situ Treatment of Contaminated Groundwater: An Inventory of Research and Field Demonstrations and Strategies for Improving Groundwater Remediation Technologies, 1992.

11

FIFTEEN GOOD REASONS NOT TO BELIEVE THE FLOW VALUES FOR A GROUNDWATER REMEDIATION DESIGN

Evan K. Nyer and David C. Schafer

INTRODUCTION

TOO OFTEN WE APPLY GROUNDWATER treatment as a separate project from the groundwater investigation and hydrogeological activities. Many groundwater treatment designs are based on the results of groundwater investigations that either fail to focus on getting the right data or yield data insufficiently reliable for treatment purposes. We simply accept the numbers that were provided from the initial investigation stage of the project. This is not good practice. In designing a treatment system, we must do more than calculate the specifications for pipes, pumps, tanks, and various equipment. In order to ensure success, we must understand where the numbers upon which we are basing our design are coming from, and what they mean. We need more than mechanical, civil, or electrical engineering. We must practice remediation engineering. Successful remediation is the result of a team of experts all working on the project together. Remediation engineers must have an understanding of the other disciplines on the team.

The problem is that as soon as I mention hydrogeology, I will probably lose most of the readers who are engineers. On the other hand, if I write an article about 15 ways that a hydrogeologist may have made mistakes, thousands (maybe hundreds) of engineers will read every word. The end result will be that the engineers who read the entire article will have a better understanding of hydrogeology. This is an "end results" kind of article.

I have asked Dave Schafer, Geraghty & Miller, Minneapolis, to help me get these important points across to the readers. Dave is one of the best teachers I have ever met, and his area of expertise is hydrology. I'm just part of the team on this project.

USING PUMPING TESTS TO DETERMINE THE FLOW FOR A GROUNDWATER TREATMENT SYSTEM

The first step in the design of any groundwater treatment system is to determine the quantity of groundwater that will be pumped from the ground. No matter what technology is used, the flow rate will be a significant factor in the detailed design of the equipment.

While pumping, groundwater cannot generally provide rapid cleanup of contamination; it does serve the purpose of hydraulically containing contaminants on-site so they will not reach downgradient receptors. If a sufficient quantity of groundwater is pumped, then all contaminated water will move toward the extraction well, while only nonimpacted water (outside the contamination plume) will migrate off-site. The hydraulic containment system can be effective only if enough groundwater is pumped to intercept the entire plume.

Determining the required extraction rate is generally a straightforward exercise once the aquifer transmissivity (T) is known. Well-known equations or simple computer models are generally used for this purpose. In all cases, the required flow rate is directly proportional to T (and is also directly proportional to the hydraulic gradient and the plume width). The key to a successful hydraulic containment system design, then, is accurately determining the transmissivity.

Transmissivity is usually determined by analyzing data from a carefully controlled constant-rate pumping test. Time-drawdown and distance-drawdown graphs are prepared for the pumped well and observation wells on either semi-log or log-log graph paper. Then, up to a dozen or more standard analysis techniques may be applied to the data to calculate the T value. Log-log plots usually require curve-matching analysis, whereas semi-log plots rely on simple straight-line analysis methods. Numerous reference texts and articles are available describing the various mathematical analysis methods that can be applied to pumping test data. (See references at the end of the article for further reading on the subjects presented in this column).

Unfortunately, real wells and aquifers rarely fit the mathematical models used to analyze the data obtained from them. As a result, data analysis is often tricky and, unless the analyst is highly experienced, can lead to errors in the T values obtained. Errors made in calculating T result directly in improper flow-rate selection for extraction wells.

To see the kinds of interpretation errors made in pumping test analysis, it is useful to review several improperly designed capture systems and the aquifer test anomaly responsible for each. Only semi-log analysis examples will be used because they provide more clarity than log-log plots. Using semi-log methods, drawdown is plotted on the linear scale with either time or distance on the logarithmic scale and T is determined from the slope of the straight line of best fit through the data points. T is inversely proportional to the observed slope, and, thus, a steep slope produces a low T, whereas a flat slope produces a high T value.

The first example illustrates the effect of recharge to the pumped aquifer. Figure 11.1 shows time-drawdown data collected from a well near a river. As can be seen in the graph, recharge from the river has caused a flattening of the time-drawdown plot, resulting in this case in a tripling of the computed T value. Fortunately, the early data from this test show the correct T value. Often, however, the early data are masked by flow rate fluctuations or casing storage effects or are simply not recorded and the recharge-affected value is interpreted to be the correct one. This would result in overestimating the required extraction rate, leading to a capture system overdesigned by a factor of three, as shown in Figure 11.2. No matter how good the groundwater treatment system design is, the remediation design is wrong. The added cost of three times the flow rate is obvious.

Another phenomenon that can affect pumping test data is casing storage in which the early data define a slope substantially steeper than the theoretical slope, as shown in Figure 11.3. This slope results from the extra time it takes to remove the water standing in the well casing just after starting the pump. Casing storage effects are particularly pronounced in wells with low specific capacities or large casing diameters. Although the form of Figure 11.3 is similar to that of Figure 11.1, the interpretation is quite different. Here, the initial slope reflects casing storage, whereas the second slope provides the correct transmissivity. The error made in this particular test is that the casing storage slope was used to calculate T, producing a value about 10 times smaller than the true value. A hydraulic containment system designed based on this T value could capture only 10% of the plume, as shown in Figure 11.4.

Another common mistake in data interpretation is failure to correct observation well drawdown data for partial penetration when using distance-drawdown graphs. In many investigations, observation wells are completed with

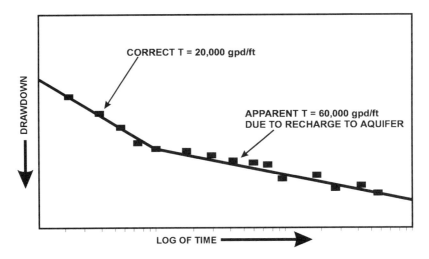

FIGURE 11.1. Time-drawdown graph showing effect of recharge on calculated transmissivity.

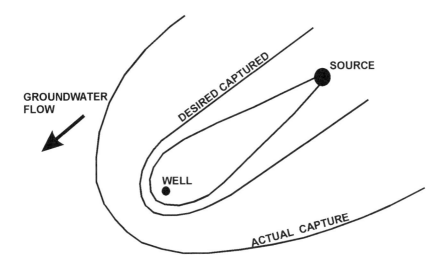

FIGURE 11.2. Capture zone resulting from overpumping by a factor of 3.

relatively short screens that penetrate only a portion of the aquifer, usually the top. A partially penetrating observation well responding to a partially penetrating pumped well will produce a different drawdown than if both wells were fully penetrating. Since most distance-drawdown analysis methods are valid only for fully penetrating wells, correction factors must be applied to drawdowns obtained from partially penetrating wells prior to analysis.

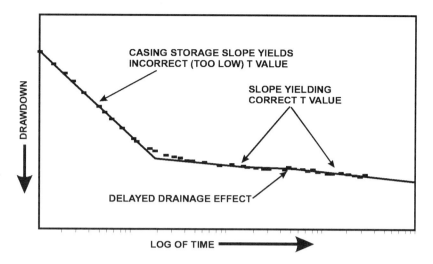

FIGURE 11.3. Time-drawdown graph showing effect of casing storage on calculated transmissivity.

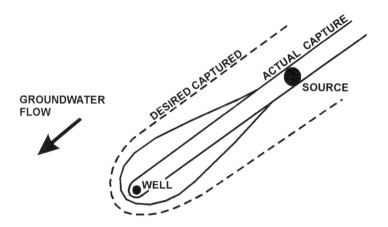

FIGURE 11.4. Capture zone resulting from underpumping by a factor of 10.

Figure 11.5 shows distance-drawdown graphs constructed from actual data and corrected data from partially penetrating wells in a highly transmissive glacial aquifer. Note that the slope produced from the actual data is about double that of the corrected data. This means that using actual data rather than corrected data would produce a T about half the true value. The resulting remediation system would capture only half of the contaminant plume.

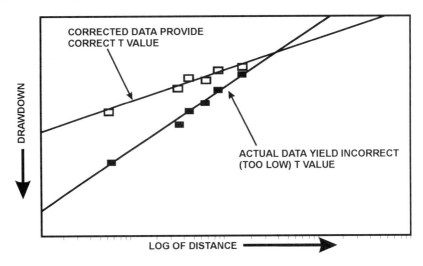

FIGURE 11.5. Comparison of actual distance-drawdown data and data corrected for partial penetration.

There are several other ways in which pumping data can be misinterpreted. This column does not have room to present a detailed analysis of each. However, in order to keep a high numeric value in the title, the following are other problems that can occur:

1. aquifer heterogeneity—an increase or decrease in transmissivity some distance from the pumped well
2. delayed drainage associated with unconfined aquifers
3. leakage from adjacent aquifers (above or below the pumped aquifer)
4. flow rate fluctuations
5. aquifer boundaries near the pumped well
6. barometric pressure effects
7. influence from nearby wells
8. precipitation during or just prior to the test
9. improperly placed observation wells
10. imprecise water-level monitoring equipment
11. insufficient flow rate.

In some cases, the anomalous behavior of the data can be managed by using a specific method of analysis or combinations of two or more methods. It is important, however, to understand the limits of validity of any method used so that it will not be extended beyond its valid range.

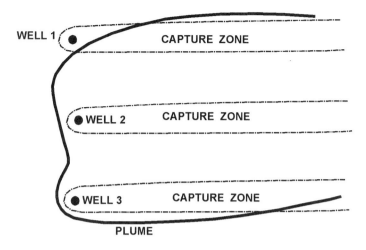

FIGURE 11.6. Capture zones show that underestimating T results in too low an extraction rate to capture plume.

WELL SPACING

ANOTHER KEY ELEMENT IN THE design of hydraulic containment systems is the proper spacing of wells if more than one well is used. If the wells are spaced too far apart, contamination can escape between wells, even though the combined flow rate has been calculated correctly.

Figure 11.6 shows a three-well system designed to capture a hydrocarbon plume. The initially proposed flow rates were insufficient to capture the plume because the transmissivity had been underestimated, largely due to failure to correct the distance-drawdown data for partial penetration. The pumping test data were reanalyzed and revised values were obtained for transmissivity and extraction rate. When the correct extraction rate was applied to the existing wells in a computer model, however, gaps remained between the capture zones of the individual wells because the wells were spaced improperly (Figure 11.7).

One solution would have been to increase the flow rates sufficiently to cause the capture zones to coalesce. The result would have been increased pumping and treatment costs for the client. Even though the gaps appear small, a 50% increase in the extraction rate would have been required to eliminate completely the leakage between wells. Once again the remediation design would have been wrong. The person paying for the resulting treatment system would not care which member of the team failed. He/she would just be stuck with a non-cost-effective remediation.

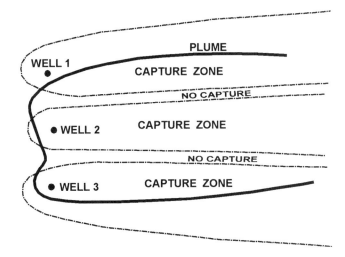

FIGURE 11.7. When the proper flow rate is produced, capture zones show that complete capture still is not achieved because wells are spaced too far apart.

SUMMARY

COST-EFFECTIVE REMEDIATION REQUIRES A team of experts. The only way that a team can work together is if all the team members have an understanding of the other areas of expertise. While this article will not provide enough information to select a flow rate, it should stimulate some intelligent questions among the team members.

The flow rate of the treatment system is too important to simply accept a number. If you are the engineer on the project, you must be able to understand the basis for the number. If you are the project manager, you have to understand how the number could be wrong and how to go about obtaining the correct one.

FURTHER READING

Dawson, K.J. and J.D. Istok. *Aquifer Testing. Design and Analysis of Pumping and Slug Tests.* Lewis Publishers, Boca Raton, FL, 1991.

Driscol, F.G. *Groundwater and Wells,* Second edition. Johnson Filtration Systems, Inc., St. Paul, MN, 1986.

Hantush, M.S. "Drawdown Around a Partially Penetrating Well." *J. Hyd. Div., Proc. of the Amer. Society of Civil Engineering,* 87, No. HY4, 1961.

Javandel, I. and C.-F. Tsang. "Capture-Zone Type Curves: A Tool for Aquifer Cleanup." *J. Ground Water,* 24(5), 1986.

Kruseman, G.P. and N.A. de Ridder. *Analysis and Evaluation of Pumping Test Data,* Second edition. International Institute for Land Reclamation and Improvement, The Netherlands, 1990.

Lohman, S.W. *Ground-Water Hydraulics.* Geological Survey Professional Paper 708, U.S. Geological Survey, Denver, CO, 1979.

Neuman, S.P. "Theory of Flow in Unconfined Aquifers Considering Delayed Response of the Water Table." *Water Resour. Res.,* 8(4), 1972.

Neuman, S.P. "Effect of Partial Penetration on Flow in Unconfined Aquifers Considering Delayed Gravity Response." *Water Resour. Res.,* 10(2), 1974.

Schafer, D.C. *Casing Storage Can Affect Pumping Test Data.* The Johnson Drillers Journal, Johnson Division, UOP Inc., January–February, 1978.

Strack, O.D.L. *Groundwater Mechanics.* Prentice Hall, Englewood Cliffs, NJ, 1989.

12

AQUIFER RESTORATION: PUMP AND TREAT AND THE ALTERNATIVES

Evan K. Nyer

I RECENTLY HAD THE PLEASURE of attending a unique seminar sponsored by the NGWA. John Cherry, of the University of Waterloo, was asked to assemble a seminar addressing the problems that have occurred with pump and treat, and the alternative technologies available to replace pump and treat. The seminar consisted mainly of two days of oral presentations and over 50 poster presentations.

The presentation was unusual because an individual made the decisions as to the seminar's content and presenters. This format had both advantages and disadvantages. One of the main advantages was that the latest thinking and data were presented by the senior member of the project team, in contrast to students, mid or junior level engineers, and hydrogeologists. This approach was very stimulating because their insights were unique, and they took their presentation to a higher standard. The disadvantage of the "latest" data was that written papers had not been prepared, and therefore, no proceedings were available for people not able to attend the seminar. This article will help solve this problem. I will provide a quick review of the seminar by highlighting the main points that were made in each session. There is not enough room in this article to review every lecture. However, each session of the seminar concentrated on one particular subject area, and the main points can be covered by reviewing a couple of lectures from each. I hope that this will provide those not able to attend the seminar with some of the information provided by the speakers. Unfortunately the seminar extended to late Friday afternoon. In order to get home for the entire weekend, I missed the last afternoon of lectures and will not be able to report on those here.

PUMP AND TREAT: PROCESSES AND LIMITATIONS

THIS SESSION SET THE STAGE for the entire seminar. We have been reading articles about the limitations of pump and treat systems for groundwater remediation since the late 1980s. Previous studies have concentrated on performance data from full-scale installations. We knew that there were limitations, but we did not know why. This seminar session provided the "why."

Three lectures concentrated on an individual reason for pumping limitations. "Remediation of Multicomponent Plumes in Heterogeneous Aquifers" by Douglas M. Mackay and E.A. Sudicky and "Hydraulics of Capture and Plume Manipulation" by John L. Wilson both concentrated on the effects of heterogeneous conditions. "Influence of DNAPL on Performance of Groundwater Pump-and-Treat Remedies" by Stan Feenstra concentrated on the solubility property of the compounds to explain pumping limitations.

Mackay explained the problem of pump and treat by the pattern of flow of water through an aquifer. If the aquifer is heterogeneous (and no aquifer was perfectly homogeneous), then the pattern of water movement through the aquifer would not be uniform. Low resistivity paths in the aquifer would carry most of the water flow. Pump and treat systems rely on water as the carrier to remove organic compounds from aquifers. The design of the pumping system is based on moving water through the contaminated zone. Mackay showed that the water would have a preference for the main flow paths and would not come into contact with the entire aquifer. He then demonstrated that even compounds that were soluble in water would still not be completely removed by a pumping system. The water would not come into contact with all of the places in the aquifer where the compound was present.

John Wilson described the problem from the other direction. He showed that the heterogeneity of the aquifer would control the removal of organic compounds due to low flow areas in the contaminated zone. The main mechanism for organic removal from these areas would be diffusion. The areas could be anything from the classic clay lens to lightly fractured bedrock. Lenses that are relatively small can still control chemical removal from the aquifer. The organic compounds will leak out of these areas for years. Any time that diffusion controls the removal of compounds from the aquifer, the cleanup time will be enormous.

Stan Feenstra described the limitations of pump and treat based on solubilities of organic compounds. He showed that even if the aquifer were perfectly homogenous, a pool of pure organic compound sitting on the bottom of the aquifer would be removed by a pump and treat system at a rate limited by the compound's solubility in water. He provided calculations that showed that it would take over 1,000 years to remove perchloroethylene (PCE) by using water as the carrier. More soluble compounds would take less time.

The remaining authors did not concentrate on only one aspect of pump and treat limitations. They showed how all of the limitations combined to prevent pump and treat from being a viable remediation method. Several authors showed that the proper use of pump and treat was plume containment.

I think that the best way to summarize all of these descriptions is simply to say that the problem with using water as the carrier to remove organics from an aquifer (pump and treat) is the natural conditions of an aquifer will continue to impede the removal of the source of contamination. In a properly designed remediation system, the first step is to remove the source. We design caps and slurry walls, remove soil, close ponds, etc., all as a first step to remediating a site. Then, as a second step, we address the aquifer. All the papers showed that it is nearly impossible to remove all of the contaminant source. Contaminants in clay lenses (and other low flow areas), fractured bedrock, NAPLs, their low solubility, etc., all continue to act as a source of contamination. Water movement through the aquifer will not address these sources.

The two conclusions that I made from all of the lectures in the first session were: (1) our technical knowledge of aquifers, organic compounds, properties, and the interaction of organic compounds with soil geochemistry all show that pump and treat should not be able to clean an aquifer, and (2) heterogeneity is probably the largest limiting factor of pump and treat remediation.

PUMP AND TREAT: CASE STUDIES

THE SECOND SESSION TOOK THIS theoretical knowledge and showed that real pump and treat systems behave as predicted. Six case histories were presented from projects that had been pumping for many years. In general, all of the project results showed rapid removal of chemicals at the beginning of pumping with a gradual decrease ending in an asymptote. All of the authors found pump and treat an excellent method to control plume movement, but were unable to remediate the site with this technique.

For practical reasons I am only going to review one paper from this session. "Long Term Performance of Pump-and-Treat Systems" by Nicholas Valkenburg reviewed two pumping systems that had been running for many

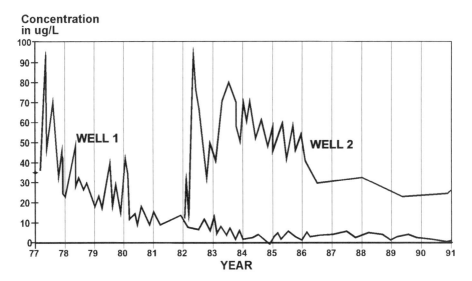

FIGURE 12.1. Carbon tetrachloride concentrations in Wells 1 and 2.

years. The first system has been pumping groundwater contaminated with carbon tetrachloride since 1977. Two interesting observations can be made from the data (Figure 12.1). First, the concentration curves for both recovery wells show the classic life-cycle concentration from a pumping system. The second observation is that there is an interaction between the two pumping systems. Well 2 was installed upgradient of Well 1. When Well 2 started pumping in 1982, there was a significant decrease in the concentration in Well 1 (Figure 12.1). Pumping Well 2 eliminated most of the source of carbon tetrachloride that was reaching Well 1. Even with this major source reduction, Well 1 did not completely "clean up," but only achieved a lower asymptote level. More pounds were removed with this dual pumping scenario, but full remediation was still not achieved.

In the second case history, chromium (Cr^{+3}) was the main contamination (Figure 12.2). The source of the chromium was an unlined surface impoundment that had been used for waste disposal. The zero line on the figure represents the closure of the impoundment. The interesting part of Figure 12.2 is that all of the monitoring wells showed a steady decrease in chrome concentration and that MCLs were reached in all of the wells by the fourth year. This was achieved without any treatment.

The difference between chrome and a chlorinated hydrocarbon is that chromium reacts with the soil of the aquifer. Carbon tetrachloride is present in the aquifer in equilibrium between soil adsorption and water solubility. This equi-

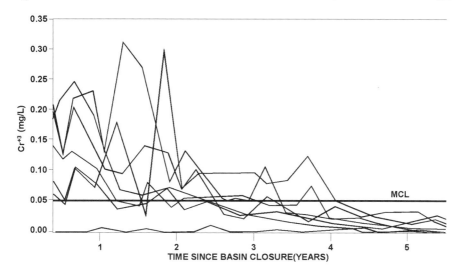

FIGURE 12.2. Chromium concentrations in shallow sand aquifer.

librium sets up a slow release of the carbon tetrachloride to the water. Chromium reacts with the soil, ion exchange, precipitation, etc., and is removed from the water. Once removed, the chromium has very little release back to the water. Without the release back to the water, the chromium (no matter how heterogeneous the aquifer) cannot act as a source. With all of the sources removed, the aquifer can achieve "clean."

CUT-OFF WALLS AND IN-SITU TREATMENT ZONES FOR DISSOLVED-PHASE MIGRATION CONTROL

THE THIRD SESSION OF THE seminar dealt with cut-off or reactive walls and in-situ treatment. Six papers were presented, with five papers covering the various walls and one covering in-situ methods. I found that there were two main problems with this session. First, almost all of the papers assumed broad economic assumptions. The assumptions went something like this: pump and treat systems were not going to clean up an aquifer; therefore, they would have to operate for extended periods of time. This extended operation obviously made pump and treat too expensive. The final leap of this assumption was that cut-off or reactive walls would obviously be less expensive. The second problem was that the entire session was devoted to theoretical methods and very little actual data was presented. In fact, very few calculations were provided to support the assumptions made during the session. All three of the reactive wall papers simply presented the concept.

The first paper was by Guy Patrick and it was entitled "Low Permeability Cut-Off Walls for Source-zone Isolation." This paper provided a good overview of the use of cut-off walls for plume control. It showed how to use these walls by themselves or in conjunction with pumping systems. Several design scenarios and case histories were presented. Personally, I wish some economic data would have been provided with the case histories to show the relative cost of pump and treat and cut-off walls.

The second paper that dealt with cut-off walls was "Hydraulic and Architectural Characteristics of Microbial Biobarriers," by M. Jim Hendry and John R. Lawrence. This paper presented a new idea for creating cut-off walls. Based upon work done in the petroleum industry, there are certain bacteria that grow large amounts of exocellular polymers when give sufficient food and nutrients. These bacteria and their polymers fill all of the void spaces in the aquifer. The idea of this paper was to use these bacteria to create cut-off walls in contaminated plumes. I am afraid that I'm going to have to be a little skeptical on this idea. Contamination plumes are extremely low in concentration compared to petroleum hydrocarbon recovery, and bacteria do not maintain exocellular polymers when the food supply is cut off. The wall would not last very long. Until further field results are presented and reviewed, I will withhold judgment.

The other three papers considered reactive walls. Reactive walls are a new idea for retaining a plume. The basic concept is to place a wall of material in the path of the plume movement. Cut-off walls are designed to stop the plume movement by stopping the entire flow of water. Reactive walls are designed to allow the water to pass through the wall, but the containments in the plume react with the material in the wall and do not pass through. The walls could treat anything from metals to nitrates. Design concepts have included biologically active walls and walls that can remove chlorinated hydrocarbons by reductive dehalogenation.

After the first paper by David Burris titled, "In-Situ Treatment Zones: General Concepts and Approaches," I thought, "Wow, what a great concept." I made a mental note to watch this technology as it developed. The second paper, "Geochemical Remediation of Ground Water by Permeable Reactive Walls: Treatment of Chromate, Nitrate, ad Halogenated Organic Chemicals" by Blowes, O'Hannesin, Ptacek, Robertson, and Gillham, provided laboratory and pilot results. After its presentation I thought, "OK, more detail is good. I definitely have to watch this technology." The third paper was supposed to be presented by Robert Borden on biologically active walls, but he was unable to attend. Instead another general paper on reactive walls was presented. My final thoughts were, "Enough, I get it! I promise I'll watch the technology, only no more papers on it today!" Now that I can look back on these presentations with calm reason, I suggest all of the readers watch the development of this technology. The only problem that I see is that I'm not sure that the public will accept contaminants simply controlled and not treated. Reactive walls will become part of the remediation, not a golden bullet; some type of in-situ treatment will still be necessary.

The final paper in this section was "In-Situ Air Sparge Systems For Plume Control," by J.F. Pankow and R.L. Johnson. This paper provided some very interesting results. Pankow and Johnson set up a large tank in their laboratory in which they ran several air sparging tests. First they filled the tank with gravel and set an air sparging system in the gravel. By controlling the water level in the tank at the top of the gravel, they were able to observe the air pattern as it hit the surface. As expected, the air pattern in the gravel was very close to the center of the well. The air did not spread out from the well as it rose through the simulated aquifer.

The experiment was repeated with sand; two interesting results were recorded. First, the air spread out from the well as it rose through the simulated aquifer. This was to be expected. Second, the air followed preferred paths as it rose up through the sand. The bubbles did not come to the surface in an even pattern. The bubbles only showed up in specific locations. Even after the air was turned off, and then turned on again, the bubbles showed up in the same location as before. This would seem to say that even though air sparging is capable of removing volatile organic compounds faster than pump and treat, this in-situ technology may face some of the same limitations when trying to completely remove organics from the aquifer. The air may not reach every location in the aquifer, and diffusion may still control the removal of the last fraction on the contaminants.

CONCLUSIONS

I PERSONALLY WANT TO THANK John Cherry for taking the time to put this seminar together. I normally expect to find 5–10% of the papers presented at a conference to have any value at all. While I have had fun with some of the presentations in this article, in general, I found 60–75% of the presentations to be very good. The good news is that this was an amazing group of senior people and presentations. The bad news is that this was Las Vegas, and I expected to take 90% of my time to have fun and to catch the remaining few good papers. I can't get rid of this feeling that Professor Cherry owes me a couple of days of vacation.

The NGWA should sponsor more seminars with this format. One seminar a year should stay away from committees deciding who and what is presented. A single person should invite senior professionals whose work and presentation style he is familiar with. While we will run the risk of too many papers on a favorite subject, the superior results clearly outweigh any possible limitations.

This seminar provided an excellent review of pump and treat remediation. I hope that this article has provided some of the insight of the seminar for those who could not attend. I am sure we will continue to see a series of papers on all of these subjects.

PART 3

WORKING WITH REGULATORS

13

THE POLITICAL DESIGN

Evan K. Nyer

INTRODUCTION

DESIGNING A REMEDIATION SYSTEM FOR groundwater or soil is not a simple process. Reading about the latest treatment technology will not prepare you to apply that technology correctly to an actual site. There are many factors that contribute to a successful, cost-effective remediation. The physical and chemical processes of a treatment technology are only one factor.

Another factor is that remediation brings several "players" to the table. My own experience is from the industrial side. These are the people who usually own, and more often pay, for the investigation and remediation of the site. However, it would be stretching it to describe this group as in control of the remediation. We also have the regulators contributing to the process. Federal, state, and local public officials, usually connected to an environmental department, bring their views and opinions to the table. The public may also play a direct role in a remediation. This can be through individuals or a community group. One final group that has a reserved seat at the table is the lawyers.

All of these groups can have a say in the final remediation design. You may have the best technical approach for a site, but if that approach does not satisfy the requirements of one of the groups, that technology may not be allowed to be applied to the site. A remediation design is not complete until it meets all of the requirements from all of the groups involved in the site. The remediation design needs an approach that includes all of the groups.

ALTERNATIVE DESIGN APPROACHES

THE FOUR MAIN DESIGN APPROACHES that I have seen employed over the years are: (1) The Technical Design, (2) The Low Bid Technical Design, (3) The Logical Design, and (4) The Political Design. All of these design methods can be used to develop projects that end up with wells, pumps, piping, treatment equipment, operation, maintenance, and spending lots of money. However, I have found that the Political Design Approach is the best way to minimize the cost of a remediation. Let us review these approaches and see what we can learn from each.

The Technical Design

I WAS TAUGHT THE TECHNICAL Design approach in engineering school. Good technical design requires a combination of the following: a thorough education in the basic sciences (math, chemistry, biochemistry, thermodynamics, etc.); well-rounded training in the various treatment methods that are being employed in the remediation field (air stripping, carbon adsorption, biological methods, in-situ methods, etc.); and experience. Experience doesn't mean simply working 10 to 20 years in the environmental field. Experience refers to getting your hands dirty installing and operating a treatment system. All of these components do not have to be supplied by the same person. Young engineers can do most of the design work as long as a senior, experienced engineer is working with them.

The technical design approach uses the latest knowledge in treatment systems, combines that with a thorough knowledge of the site, and experience with the practical applications of the equipment in the field to develop a cost-effective design. As we have discussed before in this column, the best remediation designs come from a combination of expertise (geology, geochemistry, modeling, risk assessment, etc., in addition to engineering). The technical design approach makes the best use of all of this expertise in order to come up with the correct remedial design.

Most of the readers of this column are from a technical background and probably think that this is the only approach to a remedial design. The trouble is that the other groups that are involved with a project may have a slightly different agenda, have a completely different view of what words like "cost-effective" mean, and have a different criteria for determining a successful remediation. Let us go through the other approaches, and then we will review some specific examples.

The Low Bid Technical Design

THE LOW BID APPROACH HAS all of the requirements of the standard technical approach. The main difference is that the low bid approach goes a little light on the experience content. Think of it as "Lite" Remedial Engineering. Low bid technical designs are usually done by engineers who are 2–5 years out of college. These engineers are competent, well-trained, and intelligent. However, when you leave out the experience component some strange things happen. You can usually recognize a low bid design by one or more of the following characteristics:

- The designers are overconfident—until you install a system that doesn't work as you expect, you are missing an important part of your experience.
- Redundant systems are used—when you have no experience in the actual operation of a system you have a tendency to back up the system; i.e., placing carbon adsorption behind an air stripper for treating TCE.
- Calculations are gospel—the designer actually believes that if a model states that the groundwater will reach 5 ppb benzene in five years, a time and date could be connected to that event.
- Hot technologies are selected as part of the remedy—for some reason the environmental field loves to select a technology that has been successful on a particular compound or site and apply it to all compounds at all sites. UV oxidation seems to be on its last legs as a hot technology, and I predict that air sparging will replace it soon.

The Low Bid approach has the rest of the limitations of the regular technical design approach. This is still a technology-based approach, and other criteria are not always considered when the remediation is originally designed. However, the Low Bid approach has its place in the environmental field. The best example of appropriate application of this approach would be gasoline station remediation. The redundancy of the problem and solutions makes experience less important for gasoline station remediation.

The Logical Design

THE BASIC REQUIREMENTS OF THE Logical Design approach are: a liberal arts undergraduate degree (history or political science can substitute); a graduate degree from any law school; and, at least, two months of thorough experience in the environmental field. The approach uses the premise that good, clear thinking and the reading of one technical paper can be combined to provide the basis for the optimum remedial design. The best way to summarize this description is to state that lawyers hate it when engineers practice law, and engineers hate it when lawyers design.

My favorite example of the logical design was the time when I was presenting an initial design recommendation to the full PRP group of a Superfund project. There were 57 PRPs, and they were represented by 57 separate lawyers. (It was close to a religious experience for me. I now know what purgatory feels like.) The point that I was trying to make during the presentation was that the present remedial design required by the ROD would cost the group $10,000,000 more than was necessary to clean up the site. I was trying to explain that the most important decision that the group had to make was whether to try and change the ROD. For the next two hours every lawyer in the group wanted to give me a suggestion on a different technology that should be included in our design considerations. We never discussed strategy or the $10,000,000. All they wanted to do was design the treatment system.

The Political Design

THE MOST IMPORTANT PART OF the Political Design approach is to anticipate what the other parties at the table need and design for those needs in addition to the technical requirements of the remediation. You must have confidence that you can design a more cost-effective treatment system than the other parties. If your cost-effective design includes what they need, they will accept it. If your design does not meet their requirements, then there is a good chance that you will not be allowed to apply the correct technology to that site. A good designer does not lose control of the design process.

There are several ways in which this can be accomplished: (1) know where the money is spent on a technology; (2) put the right name on the technology; (3) use the words "pilot plant" generously; and finally (4) the three most important words in any remediation design are strategy, strategy, and strategy. The best way to explain each of these ways is to use some examples from designs that I have completed.

WHERE THE MONEY IS SPENT

THE TECHNOLOGY THAT MOST PEOPLE love to hate is carbon adsorption. Everyone feels that this is an expensive technology. The problem is that carbon adsorption is an expensive technology to operate. It is not a expensive technology to purchase. The capital expense for a carbon adsorption system is relatively low.

In 1986 I had the chance to replace a carbon adsorption system that was treating a groundwater contaminated with methyl ethyl ketone, MEK. Activated carbon does not work very well on any of the ketones. Two problems were occurring with this system. First, the MEK was breaking through the 20,000 lb carbon vessel in less than two weeks. That translates into a very expensive treatment system for 25 gpm. Second, the carbon was acting as a bacteria growth media. Once the bacteria ran out of oxygen, they started using the sulfate in the groundwater. This produce the classic rotten egg smell of hydrogen sulfide. The neighbors did not appreciate this result.

I was able to convince the plant that a biological treatment system would be a better technology than the carbon adsorption. The fact that it was not going to cause any odors and that it was going to cost one-fortieth of the carbon system made the job easier. The next step was to convince the state EPA. This is where the problems were encountered. I was told that this was going to be the first biological treatment system installed on groundwater in the state of California. The regulators had no basis for reviewing the capabilities of the biological treatment system to remove MEK from the groundwater. We came very close to not being allowed to use biological technology until, out of frustration, I said that we would place carbon adsorption after the biological system to ensure that no MEK was discharged.

Two questions should come to mind at this point. First, why did I suggest a technology that I knew was 40 times more expensive than biological treatment, and second, why would the state accept a technology that was already not working at the site? The answers are the key to the Political Design Approach. Carbon absorption is expensive to operate. When the groundwater was loaded with MEK it required large quantities of activated carbon to remove the compound. However, if the biological system removed all of the MEK first, then the carbon would just sit there and not have to be replaced. An activated carbon system only cost $8,000 for the 25 gpm flow, and there was no operating cost associated with it because the biological system was already meeting the effluent criteria. Know where the money is spent.

For the second answer, you must remember that this was 1986. Very few treatment systems had been installed on groundwater at that time. All of the literature referred to activated carbon as the ultimate way in which to remove contaminants from water. The state could not rely on the industry to protect the public. In their mind, the industry would do anything to save money. In the end the data from the plant was less important than independent sources of literature. The experts said that carbon absorption would work, and the state could not be confused by the facts.

PUT THE RIGHT NAME ON THE TECHNOLOGY

I WAS WORKING ON A state Superfund project in 1990 that had soil contaminated with toluene. It was decided that in-situ biological treatment would be used to remediate the soil. The problem was that part of the project was going to be run during the rainy season, and large quantities of water were going to be generated that would be contaminated with toluene.

Figure 13.1 presents a treatment system design which would take care of the contaminated water that was generated. Figure 13.2 presents the far superior design that was able to save the client $500,000 over the two years the system operated. Figure 13.1 assumes that the toluene is a hazardous waste. It will require permits, sludge disposal, monthly sampling, and reporting. Figure 13.2 assumes that the toluene is a food source with which we could grow the bacteria needed for the soil remediation. Permits were not required, the bacteria (sludge) were returned to the soil, and the discharge was accepted by the local POTW. Figure 13.2 was accepted by the state, and installed at the site. Of course, we did not present both figures.

USE THE TERM "PILOT PLANT"

I HAVE PROBABLY USED THIS method more than any other. Any time that the regulators are uncomfortable with the approach you are taking, call it a pilot plant. The pilot plant design can be the same flow rate as the full-scale system, and you can state that the system will be left running if it performs to specifications.

The great part about this approach is that it does not require the same level of permits as a final design. In addition, regulators seem to be more comfortable when they do not think that they are committing to something permanent. Think of it as a marriage contract. More people would get married if we switched from "till death do you part" to "try for a year and see if it works."

My favorite example for this method took place in California. We met for two days with the state EPA. During that time we explained a new technology that we wanted to apply to a groundwater. After a long time we were able to convince them to consider the new technology, but the state started to require excessive monitoring wells and sampling throughout the site to ensure that there were no secondary effects from the system. The new monitoring requirements were going to cost more over the life of the project than the savings from the new technology. Once again, out of frustration, I said, "But this is only a pilot plant." Immediately the notebooks were closed and the state asked why we had not said so before. They were perfectly willing to let us try

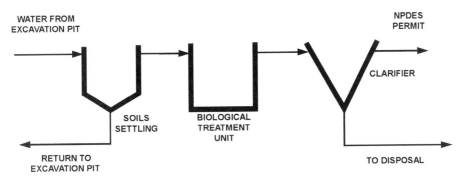

FIGURE 13.1. Biological treatment system.

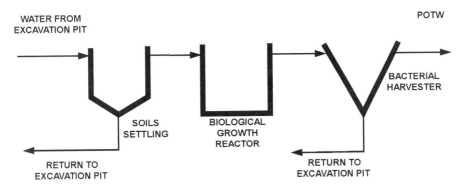

FIGURE 13.2. Biological growth reactor.

the new technology for 6 months and would expect a report at that time on its performance.

Pilot plants connote a system that will be evaluated and changed if required. The funny thing is that any full-scale system would also have to be evaluated to make sure that it was performing as designed. If flaws were found, the adjustments would be made. This is just good design practice. "Pilot plant" says this more clearly to the regulators.

STRATEGY

THE FIRST THREE EXAMPLES WERE based upon projects in which we were not confronted with alternative technologies from one of the groups at the table. While it has been fun in this article to poke fun at the other groups at the table, to be perfectly honest, they have all provided good ideas on projects. As we all get more experience in this field, the ideas will get better, or at least, harder to refute.

Modern Political Design requires that a strategy be generated for the site. Then, as the project progresses, and ideas come from all of the parties, the good ones can be incorporated. The bad ideas can be shown not to conform with the progress of the strategy. When we look at total cost of a project, the specific technology does not have as much impact as the overall strategy. For example, switching from carbon adsorption to catalytic oxidation treatment on the air stream of a VES system has less effect on the total cost than switching from in-situ methods to pump and test methods. Know your strategy and give in on minor points, especially if those points do not add significant costs.

SUMMARY

REMEDIATION ENGINEERING IS MORE THAN just putting paper and pumps together. A good design requires more than simply installing the hottest technology at the site. Remediation requires a thorough understanding of a site and a strategy based upon that knowledge, the input from the other areas of expertise, AND incorporation of the requirements from the other parties that may play a role at the site. There are always methods to clean a site and maintain control of the costs.

14

NEGOTIATING WITH REGULATORS

Evan K. Nyer

THE NORMAL WAY TO WRITE an article on negotiating with regulators is to concentrate on the regulations. There are several articles that have been written that detail the mechanisms of the various environmental laws. An excellent recent example would be the EPA's communications on determining the impracticality of reaching cleanup concentrations in the aquifer (USEPA, 1993). Everyone should review this material before initiating an aquifer remediation design, let alone a negotiation. It is important to completely understand the federal, state, and local regulations that will give the interested parties the framework under which to work toward a remediation design acceptable to all. It is equally important to establish a strategy for negotiating with regulators over remediation objectives, technologies, and costs.

Most people think of negotiations as arguing with the regulator over the proper end point of the project. When can the site be declared clean? However, my experience has been different. I have spent most of my time negotiating over the application of new technologies or changes to previously approved designs. Over the years, I have found the three factors that are critical to successful negotiation are: (1) owner history with the regulators; (2) understanding what can be accomplished at the site; and (3) recognizing that the regulator and the owner of the site have different criteria for determining a successful remediation. We will first discuss each of these factors and then review a couple of case histories to demonstrate these concepts.

OWNER HISTORY WITH THE REGULATORS

One of the most important aspects of negotiating with a regulator is the history that the owner of the site has developed with the regulator(s). I have never been able to find a method, technology, or strategy that will overcome a bad history between the site owner and the regulator(s). The first thing to remember is that the regulator must place a certain amount of trust in the property owner. If the regulator trusts the owner of the property, then the regulator will review all information and ideas presented by the owner based on technical and public merit. On the other hand, if the regulator has had a history of difficult and dishonest dealings with the owner of the site, then the regulator will tend not to believe anything that is presented by the owner.

Ninety percent of my dealings with regulators have found them open, honest, and willing to accept new ideas and technologies for the betterment of the site and the public welfare. On very rare occasions I have found regulators not willing to listen to new ideas whatsoever. In most of the cases where the regulators were not willing to listen, the blame could definitely be put on the owner of the site. My favorite experience with bad history came from a client on the east coast. This client had a tendency to take the keys out of his pocket and throw them across the table at the regulator with the exclamation "You take the damn plant." This was the client's response to any disagreement with the regulator. As you can imagine, this did not create a viable situation under which to negotiate anything. The regulator basically did not want to listen to anything that we or the owner had to say.

Probably the worse case of bad history that I ever heard of came from a project in the Midwest. The bad relationship between the owner of the site and the regulator culminated at one meeting during which the owner stood up and spat at the regulator. This should not be considered a good negotiating technique. Needless to say, we were not able to accomplish much on that situation.

The most common cause of poor relations between property owners and regulators is missed deadlines. If the property owner has a history of not responding to the regulator's requests (or demands) in a reasonable time, then the regulator will hesitate to trust the owner in the future.

Any negotiation requires trust on both sides. The first step before pursuing a negotiation with the regulator is to find out the owner's history with the regulator. Good histories will lead to fruitful negotiations, bad histories should be attacked (e.g., change the project manager that worked with the regulator in the past; have high level corporate executives apologize for past mistakes; promise a new cooperative future) before proceeding with the technical part of negotiations.

WHAT CAN BE ACCOMPLISHED AT THE SITE

THE SECOND STEP TO SUCCESSFUL negotiation is being prepared by knowing what the objectives are for your site. The objectives for any site can be formulated once you understand what can be accomplished at that site. We have discussed many times in past columns the limitations that pump and treat and in-situ technologies have when applied to a site. These limitations must be incorporated into your understanding of what can be accomplished at your site and ultimately into the site objectives. The EPA document on impracticability is an excellent reference for limitations of common and innovative technologies.

Once you have a good understanding of the site, potential remedies, and the limitations of those remedies, you can develop your strategy for negotiation. I have found that negotiation of the following three key issues can save significant amounts of money for the property owner.

- Change of remediation technology to a less costly alternative;
- Set the end point of the remediation to a level that can be accomplished given limitation(s) of available technologies [i.e., achievement of concentration asymptote(s) as an end point for the remediation];
- Use a monitoring program that reflects what is truly going on at the site and saves money over the years of operation.

Remember, it is important to have a specific understanding of which aspects of remediation are costly to property owners and which aspects create little economic impact. The negotiation strategy should focus on lessening the economic impact to the owner rather than trying to win every minor issue negotiated. Items that are important to the regulator should be evaluated based on cost to the owner, and accepted if it is determined these items can be implemented at a reasonable cost. Do not get caught with the dead-end thinking that everything must be technically correct according to you.

RECOGNIZE THE DIFFERENT GOALS OF
OWNER AND THE REGULATOR

THE OTHER IMPORTANT ITEM TO keep in mind during a negotiation is that the interests of the regulator and interests of the owner are not the same. While both parties are interested in protecting human health and the environment, other objectives will not be the same. It has been my experience that most regulators regard time as an important function (the less time the remediation takes, the better) and they prefer destruction processes over simple removal or fixation processes. Conversely, most owners concentrate on the financial burden that the remediation will cause. Most people look at this dichotomy as a bad thing. How are we to communicate over the issue when the two parties want to accomplish different goals? My view is that the different goals are the key to a successful negotiation. It is possible to develop remediation designs that provide each party with their objectives. Give the regulator a remediation that is quick and/or uses a destructive method and they will not care what the associated costs are.

In fact, it has been my experience that if the remedial design meets the regulator's objectives (fast and/or destructive) then the regulator will accept new technologies, changes in design (even from signed Records of Decision [ROD]), and cost-effective monitoring programs. These items can also achieve the owner's goal of a low-cost solution. The objective of the remedial designer should be to come up with an alternative remediation that gives the regulator(s) what they want at a reasonable cost to the owner.

The best way to understand each of these negotiating tips is to review the following case histories.

CASE HISTORIES

Seymour Superfund Site

THE SEYMOUR SUPERFUND SITE IS located in Seymour, Indiana. Originally, the Seymour Superfund Site was a chemical recycling facility at which the owner mainly stored chemicals and drums as opposed to processing them. It was one of the first Superfund sites that went through the process of investigation, design, and installation of a remediation system. The original design had a phased approach for installation of the remediation.

At the time of the negotiation, the next phase in the remediation was the installation of two wells at the northern boundary of the site in order to increase the flow and removal of contaminants in groundwater below the site. This phase included the installation of a treatment system consisting of air stripping, carbon adsorption, and discharge to a POTW. The estimated cost of this work was $1.5 million.

Most of the compounds lost to the groundwater were volatile and biodegradable. The original pump and treat system for the plume had been operational for four years prior to the design and installation of this next phase. The PRPs requested that we develop a negotiating strategy that would allow them to save the $1.5 million cost of that phase. The PRPs felt that the initial phases of remediation at the site—which included removal of the drums, removal of contaminated buildings and debris, installation of a 14-acre cap, installation of interception wells along the plume with a pump and treat system for the groundwater—had successfully reduced/removed risk to humans and the environment. Furthermore, the PRPs felt further expenditures would not add anything to the remediation. The problem was that the two new wells and the pump and treat system were requirements of the original ROD.

The keys to successful negotiations for this site were: (1) the history of the PRP group with the state and federal regulators was excellent, and (2) there were four years of monitoring data available for establishing a complete understanding of the site conditions. From the start of the project, the PRP group had been very aggressive in proceeding with this Superfund remediation. At the time of this new phase of the project, the site was two years ahead of the original schedule. This provided for a great working relationship with the regulators.

The site data also provided excellent detailed information upon which to base a technical discussion with the regulators. When the project was originally conceived and installed, it was assumed that the main factors affecting the organic movement in the groundwater would be advection and retardation. After four years of operation, the data analysis showed that this assumption was incorrect. Figures 14.1 and Figure 14.2 show the results of the plume movement for benzene and phenol at the site. The slanted line shows where the plumes were predicted to be at this time, based upon advection and retardation. The crossed lines show the actual position of these components of the plume. As can be seen, the benzene and phenol have not moved off the site.

In addition to these data, other data collected at the site included the environmental conditions in the subsurface, oxygen concentration, redox potential, presence of bacteria that could degrade the compounds in the plume, and the total organic carbon content of the soil. Based upon all of this information, we were able to conclude that natural biochemical reactions were removing the organic compounds as fast as they could move through the aquifer. (This was one of the first projects at which we started to understand the natural biochemical reactions that were occurring below ground.)

With the wealth of data collected, we were able to show the regulators that a natural destruction process was already occurring that was not considered in the original ROD. The design being discussed would pump the water from below the site to an air stripper. This process would have short-circuited the natural destruction process and resulted in organic compounds being released to the environment.

Based on the history of the PRPs with the regulators, our detailed understanding of the natural biochemical reactions that were occurring at the site, and a destruction process versus a process that would release the organics to the environment, we were able to change the next phase of the project. In this case, the regulators accepted that the natural destruction process would be

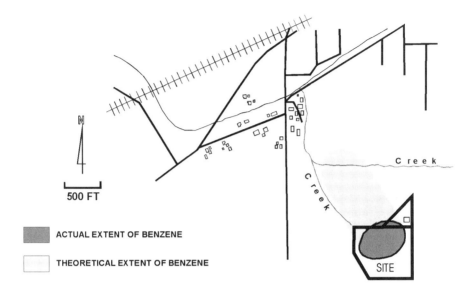

FIGURE 14.1. Theoretical and actual distributions of benzene in the shallow aquifer, Seymour site, Seymour, Indiana.

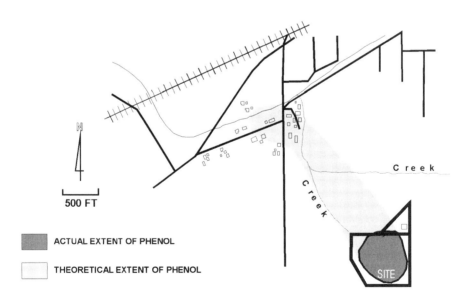

FIGURE 14.2. Theoretical and actual distributions of phenol in the shallow aquifer, Seymour site, Seymour, Indiana.

superior to the installation of two new wells with pump and treat. The savings to the PRP group was the $1.5 million minus the $200,000 required for studies to prove the destruction of the compounds.

Superfund Site in the North

THE NEXT CASE HISTORY IS of a new site that is undergoing the installation of the remediation at the present time. This is a small Superfund site in the northern part of the country. The original site was a private property. The owner brought in waste solvents to be used as a fuel source. One of the methods used by the owner was to pour the solvents over a used car, light the solvent in order to burn off all the material in the car and subsequently recover the metals from the car. While the number of drums was relatively small, the material had reached a drinking water aquifer and affected several homeowners in the area.

Figure 14.3 shows a rough sketch of the site. The site was on top of a hill and a groundwater divide. The upper water-bearing formation did not have groundwater flowing across the location. The site was made up of upper till, lower till, and bedrock. The upper till was 20 to 25 feet below surface, the lower till was 15 to 20 feet thick and then the bedrock. The upper till had a hydraulic conductivity of 10^{-5} cm/sec and the lower till was 10^{-7} to 10^{-8} cm/sec. The natural tightness of the geology had prevented a wide distribution of the original spill. Most of the compounds were still on the site, with small amounts reaching the drinking water aquifer in the bedrock.

Figure 14.4 shows the original design proposed for this site. The idea was to remove the organic compounds with a pump and treat system, treat the water above ground, and recharge the treated water to the upper till. The recharge would help flush the compounds and increase the rate of remediation. The pump and treat was to remove most of the contaminant mass with residual contamination being removed by natural biochemical reactions. The original investigation had shown that biological activity at the site was already significant, and one of the main reasons that the plume had not moved off-site was natural biochemical destruction of the compounds.

The problem with this design was mainly time. Due to the tight nature of the upper till geology, it was anticipated that the pump and treat system would have to run for 20 years. At the end of 20 years most of the mass would be removed but intrinsic biochemical reactions would still be required to reach the final cleanup goals at the entire site. The predicted cost of the entire project was $25 to 30 million.

The PRP group requested that a new design be developed that would significantly reduce the cost. Figure 14.5 shows the new design for the site. A

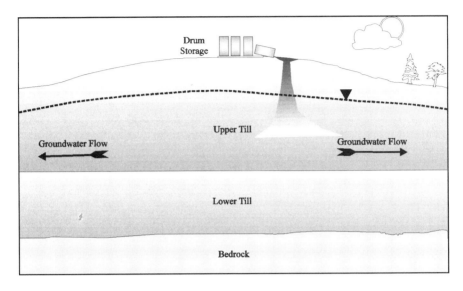

FIGURE 14.3. Superfund site in north.

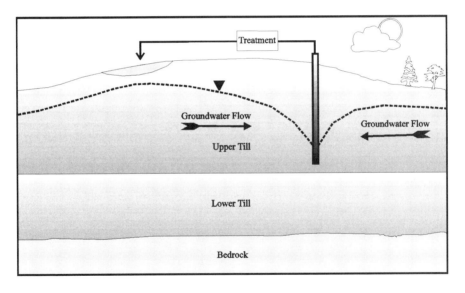

FIGURE 14.4. Pump and treat with recharge.

vacuum-enhanced recovery (VER) system was suggested for the site. Under this design, water was no longer used as the carrier to remove the compounds. Air is the carrier in a VER system. The investigation shows that most of the compounds were volatile.

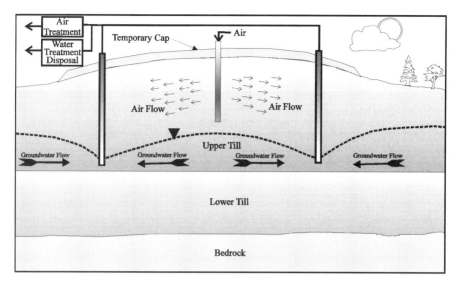

FIGURE 14.5. Dewatering with air carrier treatment.

The design thought process was as follows. Since the main source of water in the upper till was rainwater and the upper till was very tight, it was decided that it could be dewatered. A temporary cap was suggested to prevent surface water intrusion, and a high vacuum system for the simultaneous pumping of water and air across the site. It was estimated that it would take two to three months to completely dewater the site, and that once the dewatering had occurred, that the air would be able to carry out the same amount of organics in six months to one year that the water from the pump and treat system would have carried out in 20 years.

Based on this design, it was estimated that the biological portion of the project could be started in one to two years as opposed to waiting the 20 years for the original design. Both the regulators and the PRP group were happy with this design. The time for remediation would be reduced from 20 to 2 years with no increase in exposure to the public. In fact, the exposure to the public and the environment should be reduced due to the time reduction of remediation going on at the site. The PRPs were happy in that the reduced time and switch of technologies resulted in an estimated cost of $5 million for the project, a savings of $20 to 25 million over the original design.

CONCLUSION

BOTH OF THESE CASE HISTORIES demonstrate successful negotiation. The regulators accomplished their objectives of protecting human and ecological health and achieved the results in minimum time. The owners also accomplished their objectives of protecting human and ecological health, while at the same time minimizing financial impact. The keys to the happy endings at these two sites were:

1. selecting technologies and designs that could satisfy both parties;
2. understanding what could be accomplished at the site and the important natural mechanism(s) that would allow and/or limit those accomplishments; and finally
3. a trusting relationship between the regulators and the owners based upon a history of good interaction.

None of these ideas are going to win a Nobel prize, but it is amazing how much money they can save our clients. This is especially true with many of our Superfund projects. A lot of the original work that went into the development of the RODs is five to seven years old. We have learned many new things about remediation in the last seven years. Negotiations based upon our new understandings may provide the regulator with a better remediation and the PRP group with substantial savings.

REFERENCE

1. Guidance for Evaluating the Technical Impracticability of Groundwater Restoration. Office of Solid Waste and Emergency Response. Publication 9234.2-25 EPA/540-R-93-080 PB93-96350) September 1993.

15

IS THIS SITE CONTAMINATED?

Evan K. Nyer and Charles D. Senz

One of the hardest questions that we face is, "Is this site contaminated?" We ask this question in many ways and at many different times during a project. The entire reason that we run a Phase I study is to determine if the site is contaminated. The main criterion that we use to shut down the project at the end is whether the site has reached "clean." We ask the question many times during the project in order to go from one type of treatment to another, and to stop treatment at one section of the project while the rest of the project continues.

The question is also different depending on who is asking the question. When the responsible party (defined as the company, agency, or group paying for the project) asks the question it really means, "Will this cost us anything?" or "When can I stop paying?". When a Risk Assessor asks the question they mean, "Will the contamination reach anyone and cause harm?" When attorneys ask the question, they mean, (Sorry, none of my quotes got through the editors). When a state or federal regulator asks the question, they mean, "Does the method used to determine the conditions at the site follow the prescribed method delineated in the official 1,328 page regulation?" When I (or any other consultant) ask the question, I should be asking, "What steps can I take to protect human health and the environment, while causing the least financial injury to my client?"

One of the things that one notices in all of these questions is that no one ever really asks if there are organic or inorganic hazardous compounds at the site. The reason that this question is never asked is that it cannot be answered. I can imagine the look in your eyes that indicates you do not believe me. Let us go through several examples that will make my point. I have asked one of my coworkers, Charlie Senz, to assist me with these descriptions. At the end of the following three contamination scenarios, we will return to the question of, "Is this site contaminated?"

Scenario 1: Is This Site Contaminated?

IN 1934, 50 GALLONS OF gasoline were released into the soil at an 80-acre site in what was at the time a rural/agricultural area. The incident occurred when an aboveground tank of the fuel was upset by a runaway tractor.

In 1994, a developer is evaluating the purchase of the property, which now borders on the fringe of an expanding city. The developer would like to use the site for the construction of condominiums.

The contamination has degraded appreciably, as one would suspect, in the last 60 years. However, sufficient residual hydrocarbon remains within a portion of the saturated zone that if one had the dumb luck to drill into it and collect a sample for analysis, the concentration of total petroleum hydrocarbons would exceed current cleanup levels based on underground storage tank (UST) regulations in this state.

The developer retains the services of a local environmental consulting firm to conduct a Phase I Environmental Site Assessment (ESA). The consultant adheres to industry-standard methods for ESAs. The consultant looks at historic aerial photographs and sees nothing out of the ordinary. A 50-year title search indicates that the land had previously been used for dairy farming, giving the consultant no indication that any subsurface impacts may have occurred. By these standard methods, the contamination is not discovered.

The developer is somewhat cautious, having had a bad experience with another property, which at one time had been leased to a dry-cleaning business. The developer asks the consultant if there are any methods that could be used to determine if the site is clean, mentioning that a friend had told him that a relatively inexpensive soil-gas survey would have detected the problem at the last property he purchased.

The consultant subcontracts a soil-gas survey of the property. Twenty sampling locations are selected in the area of the garage and barn, because the consultant feels that this are might reasonably be expected to have experienced some sort of release. The consultant's subcontractor collects the soil gas samples through a steel probe at a depth of 5 feet below land surface. One of the samples is collected less than 50 feet from where the release occurred in 1934. The samples are analyzed for volatile organics and no contamination is detected.

In this case, the answer to our question is "No, this site is not contaminated, based on the results of industry-standard investigative practices."

Scenario 2: Is This Site Contaminated?

THE SAME CONSULTANT WAS HIRED to oversee the excavation of a gasoline UST from the site of a former cab company on the other side of town. The 1,000-gallon tank was installed above a fractured claystone.

The claystone at the depth of the UST was highly weathered, although it was fairly competent at depth. Upon removal, it was obvious that the UST had leaked. A silver-dollar-size hole was observed in the bottom of the UST, directly below the fill port. In addition, the weathered claystone was stained, and headspace analyses of the weathered claystone conducted by the field geologist indicated the presence of volatile fractions in the weathered rock.

The geologist directed the backhoe operator to remove two feet of the weathered rock from the base of the excavation. After the soil was removed, the geologist collected three samples from the base of the deepened excavation; the minimum number of samples required by the governing regulatory agency in this state. These samples were clean. What the geologist did not notice was an area of fracturing in the claystone in the north end of the excavation, near the hole in the tank. The action of the backhoe had smeared the weathered claystone, thereby hiding the fractures.

Although the three samples collected from the base of the excavation were clean, approximately 50 gallons of gasoline were present within the saturated zone both adsorbed to the subsurface materials and dissolved in groundwater.

Because the tank was known to have leaked, the same regulatory agency required that depth to groundwater, groundwater flow direction, and groundwater quality be determined. The consultant installed and sampled three monitoring wells as shown in Figure 15.1. The consultant had made a preliminary determination of groundwater flow direction based on topography. The two downgradient wells were placed at the southern property boundary, about 150 feet from the UST, and the upgradient well was installed at the northern property boundary, approximately 40 feet from the UST location. During drilling, the field geologist noted saturated conditions within some interbedded silts and fine sands at a depth of 20 feet below land surface. After evaluating the water-level data that was collected from the wells, it was determined that groundwater flow was to the south-southeast. Analytical results indicated that the groundwater was clean, and the regulatory agency issued a letter of closure.

Although geologic materials and groundwater near the UST had been impacted, the monitoring wells showed no contamination. Degradation, both biological and chemical, a low groundwater velocity, and retardation factors affecting migration had prevented detectable levels of hydrocarbon constituents from reaching the two downgradient wells. These same processes

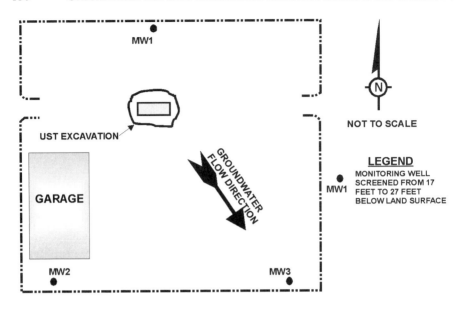

FIGURE 15.1. Site map, Scenario 1.

would continue to prevent migration of the contaminants beyond their current extent.

In this case, the answer to our question is "No, this site is not contaminated, based on the results of industry-standard investigative practices."

Scenario 3: Is This Site Contaminated?

IN YET ANOTHER SECTION OF the same town, two leaking USTs are removed from a former service station site by another consultant. In this case, the USTs had been installed above an alluvial aquifer consisting of fairly uniform coarse-grained sand. Over time, the USTs had released 50 gallons of gasoline.

The consultant advised the property owner that the most timely and cost-effective manner in which to address the contaminated soil and groundwater was to install and operate a combined air sparging/vapor extraction system (AS/VES). The consultant also indicated that three wells, one upgradient and two downgradient, should be installed and sampled quarterly during the operation of the AS/VES.

The monitoring wells were installed in an orientation similar to that described previously for Scenario 2 (Figure 15.2). Groundwater was present under water-table conditions, approximately 25 feet below land surface. The wells were sampled at the time of system startup, and all samples were deter-

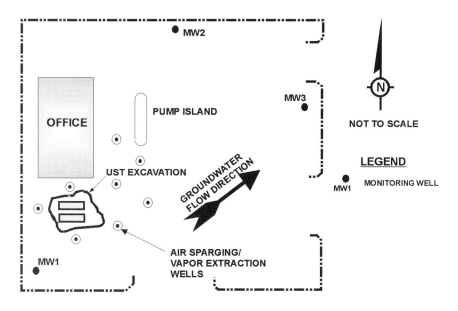

FIGURE 15.2. Site map, Scenario 2.

mined to be clean. Nonetheless, sampling and operation of the AS/VES continued for a period of one year, by which time the concentrations of volatile fraction constituents in the off-gas of the VES had stabilized at a low concentrations.

The consultant and property owner petitioned the regulatory agency for approval of closure, which was granted with the condition of an additional year of quarterly groundwater sampling. Analytical results for the four subsequent sampling events indicated no impacts to groundwater at the points of sampling. The regulatory agency granted a letter of closure.

Soil and groundwater in the immediate vicinity of the USTs still contained residual and dissolved hydrocarbon constituents. The AS/VES had been largely successful; however, some hydrocarbon impacted soil and groundwater remained in place after closure.

In this case, the answer to our question is again "No, this site is not contaminated."

When Is a Site Contaminated?

As can be seen by these three scenarios, knowing that there are contaminants present at a site and collecting the data that shows that there are contaminants at the site are two very different functions. While we took the time in each of the scenarios to describe exactly what contaminants were present, there is no way that a consultant coming on the site would be able to know where and if there were contaminants on the site. All three scenarios used the best available sampling and survey technology to find contaminants. In each case, these techniques showed that there were no measurable contaminants at the points of monitoring. We would take these results on any project and declare that the site was not contaminated or had been cleaned and was no longer contaminated.

However, in these scenarios we also have "perfect" knowledge along with our sampling knowledge. In addition to the question, "Is this site clean?" we must also ask, "Why isn't our data accurate?" One way to determine the accuracy of our data is to revisit the questions that our various parties asked at the beginning of the article. The responsible parties would be happy with our monitoring results because they could stop spending money. However, if they had "perfect" knowledge, they would be unhappy and worried about potential future liability. The Risk Assessor would be happy with the monitoring results because it shows that none of the chemicals has a pathway to reach human beings and no one is being harmed. In fact, the risk assessor is the only party that would be happy with "perfect" knowledge. There is nothing in our "perfect" knowledge that leads to a different conclusion. No one is being harmed at any of these sites. The federal or state regulator would be satisfied with the monitoring results of our three scenarios because the proper methods were followed. However, if they had "perfect" knowledge, they would definitely not be happy. They could interpret their responsibility as not protecting the public as long as contaminants are on the site. (I will not review the lawyers and my position, because lawyers only care about opinion, not knowledge, and people always get mad when I use the word perfect in any sentence that deals with one of my opinions.)

I think that this all leads to the conclusion that maybe the risk assessor has the only realistic view of determining if a site is clean. Without "perfect" knowledge, we really do not know what is present at a site. We only know what has been carried to our monitoring points by air or water. Reality is that we can never know absolutely what is at a site, we can only make sure that human health and the environment are not being harmed. This definitely calls for a group hug of your local risk assessor, and maybe a little more respect.

PART IV

BIOLOGICAL REMEDIATION

16

HYDROGEOLOGISTS SHOULD MANAGE ENHANCED BIOLOGICAL IN-SITU REMEDIATIONS

Evan K. Nyer

INTRODUCTION

ONE OF THE UNUSUAL CIRCUMSTANCES that I have noted in the field of remediation is that we rely on microbiologists to manage in-situ bioremediations. Although microbiologists are an important part of the project, I don't understand why they should be responsible for making most of the important decisions. An in-situ bioremediation project relies on combining the expertise from several disciplines in order to develop a successful project. This article will review the details necessary to understand the decisions that must be made during the design and operation of an in-situ bioremediation project.

An in-situ bioremediation project is based upon biochemical reactions occurring in a geological setting. Although bacteria are responsible for the biochemical reactions, their natural degradation rate is limited by chemical and physical factors. An in-situ bioremediation design requires the identification of the rate-limiting factors of the bacteria and the delivery of those factors to the bacteria.

WHAT ARE THE MAIN COMPONENTS OF BIOCHEMICAL REACTION RATES IN AN IN-SITU PROJECT?

AN IN-SITU BIOREMEDIATION PROJECT is made up of four major components: microorganisms, oxygen, nutrients, and environment.

The microorganisms are the workhorse of the project. The bacteria use the organics that were released to the environment as a source of food. Chemical bonds in the organic molecules act as the source of energy for the endemic bacteria and as building blocks for reproduction. Bacterial growth and reproduction occur naturally, not because they comprehend the regulatory consequences of a contaminant plume.

The large amount of time and money that has been spent on in-situ projects has gone toward trying to determine whether the bacteria that are necessary for the degradation are present at the site or if specialized bacteria must be imported. The most expensive approach to answer this question is to try to identify the bacterial species that can degrade a specific compound found at the site. In reality, multiple microorganisms work in concert during the degradation of an organic compound. In the field, a single bacterial species is never responsible for site remediation.

If the compounds are degradable, then the natural bacteria at the site are usually able to degrade the compounds. The only times that bacteria need to be introduced to a site is when a toxic condition has existed at the site and has killed all of the natural bacteria. After the toxic conditions have been neutralized, bacteria may have to be reintroduced to the site in order to speed the cleanup. New spills of degradable material can also be toxic to large portions of the natural bacteria. The cleanup of new spills may be enhanced by introducing mixed bacterial cultures.

These circumstances assume that the organics are degradable. In general, petroleum hydrocarbons are degradable and chlorinated hydrocarbons are less degradable. The more chlorine substitutions on the organic compound, the less degradable the compound.

There are many research projects that are evaluating the application of specialized bacteria for hard-to-degrade organic compounds. However, no one has successfully demonstrated the advantages of adding a specialized bacteria to a site and published the detailed information. The USEPA conducted a thorough study of bacterial additives for the Alaska oil spill cleanup. None of the products produced significant improvement when applied in the field. Many other research programs are ongoing for the degradation of chlorinated hydrocarbons and PNAs. Even when these products have been developed, delivery of the bacteria to the contaminated zone will still limit use in an in-situ bioremediation project.

While the bacteria are the key to bioremediation, at the present time we cannot affect whether the appropriate bacteria are present at the site of the organic contamination. In most cases, if the compound is degradable, the natural population has already adapted to the available organic compounds and is using the compounds as a food source. Simple microbial tests can be conducted on the soil to confirm the presence of viable bacterial populations and those that are capable of degrading the specific organic compound that was released. A microbiologist should be used to conduct the testing of soil and aquifer samples to confirm the presence of the appropriate bacterial populations.

The real object of an in-situ project is to enhance natural bacterial growth and reproduction. We do this by supplying the factor that is limiting the reaction rate of the bacteria. The main limitations are oxygen, moisture, and nutrients, NH_3, and PO_4. We must also ensure that the environment is suitable.

Oxygen is the main rate-limiting factor in organic chemical degradation. The bacteria need large amounts of oxygen to produce energy. Many in-situ projects have required oxygen without nutrient addition. However, none to my knowledge have excluded oxygen and only required nutrient addition.

Moisture is the second most important factor. Of course, moisture is only an important factor in the unsaturated zone and not the aquifer. If the unsaturated zone contains too little moisture, then the bacteria will not have the micro environments that they need to survive. Too much moisture will inhibit oxygen transfer. Laboratory tests can be run to determine the effect of moisture content. To monitor this factor on a full-scale system, humidity measurements can be made on the effluent air stream of a VES/Bio system. This will evaluate the moisture content of the soil.

The bacteria also require macronutrients and micronutrients to reproduce. The macronutrients are nitrogen in the reduced form, NH_3, and phosphorous in the most oxidized form (PO_4). Micronutrients are almost always present in either the soil or aquifer and do not have to be considered in an in-situ project. Nutrients are needed in situations where there is a need to grow a large bacterial population. This would be appropriate for large spills or when it is necessary to minimize the total project time. Both macro- and micronutrients can interact with the soil matrix. Before a nutrient delivery system is designed, a geochemist should conduct a series of tests to ensure that the nutrients can migrate through the soil and aquifer.

The final factor that must be considered is the environment. Bacteria grow best under certain temperature and pH conditions. In general, the higher the temperature, the faster the bacteria will reproduce. Although aquifers maintain a relatively constant temperature during the entire year, the best time to start up an in-situ project is the warmer months. The aquifer will maintain activity all year, but the unsaturated zone may shut down biological activity in the winter months in the northern states. pH should be kept between 6 and 8.5.

All of these factors have one thing in common; they must all be delivered to the contaminated zone. As stated above, the biological reactions are probably already occurring at the site. The main challenge in an in-situ design is the enhancement of the natural ongoing reactions. Although laboratory tests can be used to determine whether oxygen, moisture, and nutrients will increase the rate of reaction, the real design problem is how to deliver these factors to the bacteria at the contaminated zone.

In order to understand why the delivery of these materials is difficult, we need to have a better understanding of the relationship of the organics to the aquifer and unsaturated zone. We first must understand why in-situ is an important process in the first place.

WHY DO WE USE IN-SITU TREATMENT?

THE BEST ANSWER TO THIS question is that the only way that we can actually clean up an aquifer is by biological in-situ treatment. This column has discussed the limitation of pump and treat systems many times. Other authors have now published several articles which discuss that pump and treat is not capable of remediating an aquifer. Many people have realized that while you pump to control plume movement, you actually use in-situ to remediate the plume. Of course, the first time that I heard this was 10 years ago by Dick Raymond.

The primary reason that we cannot use pumping systems to remediate an aquifer is that the aquifer does not release all of the contaminants at the same time. Figure 16.1 shows the life cycle concentrations during a remediation. At first, the contaminants quickly leave the aquifer with the pumped water. However, as the pumping continues, less and less mass of material is removed from the aquifer. The concentration almost stops decreasing near the end of the project. The flattening of the line is described as the asymptote, and many projects try to use this flat area to define the end of the project. The problem with using the asymptote is that if the pumping stops, then the concentration can increase in the aquifer. The same pattern holds for the unsaturated zone and a vapor extraction system.

There are several ways to describe why the concentration stops decreasing and increases if the pumping stops. One of the best visual methods come from Szecsody (Szecsody and Bales, 1989). Figure 16.2 shows the four main mechanisms that control the rate of release of organic compounds into the water within the aquifer. (I would suggest that the reader review Keely's ar-

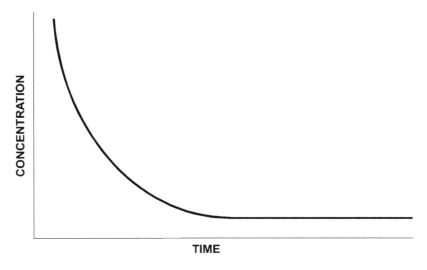

FIGURE 16.1. The change in concentration over the life of a remediation.

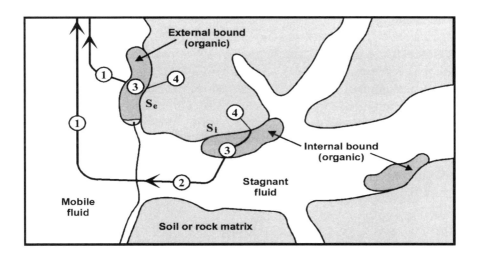

FIGURE 16.2. Transportation of contaminants in an aquifer. Conceptual model of sorption in porous aggregate. (1) Solute transport in the mobile fluid and stagnant boundary layer. (2) Intraparticle diffusion in stagnant fluid. (3) Transport in bound organic phase. (4) Binding and release within the bound organic phase and/or at the mineral surface. From Szecsody, J.E. and R.C. Bales. *J. Contam. Hydrogeol.*, 4, 181–203, 1989.

ticle (Keely, 1989) for its discussion on the relationship between solubility, diffusion, advection, and desorption on the concentration of contaminants in the water of an aquifer.) As can be seen in Figure 16.2, not all of the organic material is in contact with water that is moving. The organics in contact with stagnant (or relatively stagnant) water must overcome simple diffusion in order to be removed from the aquifer.

We can now see the relationship between Figures 14.1 and 14.2. The beginning of the project produces high concentrations in the pumped water due to the organics in direct contact with flowing water being quickly removed from the aquifer. The concentrations decrease as those organics are removed. The end of the project has a relatively constant concentration of organic compounds due to the slow diffusion of organics from stagnant water into the flowing water.

We can also explain why the concentration increases if the pumping system is turned off. When the pumps are turned off, the flowing water returns to its natural, slow movement. The organics in the stagnant water continue to diffuse into the slower moving water. The mass of organics that diffuses is the same, but the amount of water that passes through that area of the aquifer decreases. The same mass of organics in less water creates a higher concentration. The concentration increases until a new equilibrium is established between the diffusing organics and the slower moving water. When a water sample is taken from a monitoring well, the concentration reflects the new equilibrium of the aquifer.

The final point of this section is the main point of the article. The key to a successful design of a pumping system is to minimize the microcosms that are in stagnant or relatively stagnant water flow conditions. A groundwater recovery design must maximize the flow across the part of the aquifer that is contaminated. The wells and recharge system must set up a pattern of flow that accomplishes this objective. The knowledge and experience of a hydrogeologist is required to meet this performance criteria.

HOW DO WE DELIVER THE NECESSARY MATERIAL TO THE BACTERIA?

NOW LET'S TURN THE DISCUSSION around and instead of removing material from the aquifer, let us look at the processes involved in delivering material to the microenvironments in the aquifer.

Figure 16.2 shows us where the organics are located in a contaminated aquifer. We can assume that bacteria exist throughout the entire microcosm. Bacteria are present in the flowing region and in the stagnant zones.

These bacteria will interact with the organics present. As stated above, they will use the organics as a food and energy source. The rate at which they use these organics is limited by oxygen, moisture (in the unsaturated zone), and nutrient content of the microcosm. In order to increase the rate of bacterial reaction and reduce the time for remediation, the rate-limiting factors must be delivered to the microcosm where the organics and the bacteria are already present.

Now we have the same problem that existed when we tried to design a removal system. We have flowing water areas and stagnant areas in the aquifer. The flowing water areas can have a direct delivery of the growth-limiting factors. However, the stagnant areas require that the oxygen and nutrients diffuse to the specific sites. Only by diffusion can we deliver the oxygen and nutrients to the site where the biological reaction is taking place in the stagnant zones.

In a removal design, minimizing the stagnant areas was important. In an in-situ design, minimizing the stagnant areas is critical. In fact, the success or failure of the system is based upon the ability of the system to deliver the enhancement factors to the locations that need them. Since diffusion is a relatively slow process, we must minimize the area of the aquifer that relies on it. Nothing will affect the time for cleanup more than the delivery of the enhancement factors.

WHO SHOULD RUN AN IN-SITU PROJECT?

As CAN BE SEEN IN this discussion, many experts are required for the proper design of an in-situ remediation. Areas of expertise that are needed include:

Microbiologists—to ensure that the proper bacteria are present at the site, that there are no toxic conditions at the site, and to advise as to what material is required to enhance the rate of reaction of the bacteria;

Geochemists—to ensure that the material that must be delivered to the bacteria will not interact with the soil particles in the aquifer and unsaturated zone;

Hydrogeologists/Geologists—to first develop the data necessary to define the extent of contamination, and then to determine the best way to deliver the required material to the bacteria that will enhance their rate of reaction;

Engineers—to design the aboveground equipment necessary to deliver the material.

While all of these experts are required as part of the project, the question remains whose expertise is the most important, and therefore, who should make the final decisions during the project. The methods for delivering the required oxygen, moisture, and nutrients are the critical portion of the in-situ remediation. The hydrogeologist is the expert who must develop the system that will deliver these factors to the bacteria.

Hydrogeologists are the best choice for running an in-situ remediation project They have the proper education, experience, and understanding to make the critical design and operation decisions during a remediation project. All of the failures that I have seen with in-situ programs were the result of the geology or hydrogeology of a site. I have never seen the bacteria (or lack of bacteria) be the cause of an unsuccessful in-situ project. If bacteria are the reason that an in-situ project will not work, this fact can be discovered early in the project with simple laboratory tests.

Once again, all of the above expertise is required to have a successful in-situ project. But the critical points of designs and operations are the delivery of enhancing material. A hydrogeologist has the best chance of designing and running a successful remediation.

REFERENCES

1. Szecsody, J.E. and R.C. Bales. *J. Contam. Hydrogeol.*, 4:181–203, 1989.
2. Keely, J.F. Performance Evaluation of Pump and Treat Remediations. October 1989, EPA Ground Water Issue; EPA/540/4-89/005.

17

IN-SITU REACTIVE ZONES

Evan K. Nyer and Suthan Suthersan

INTRODUCTION

ONE OF THE RECURRING THEMES that I have had in this column is the concept of creating the necessary environment below ground to stimulate the reaction that is desired. A new technological area is quickly developing in the remediation field based upon this concept. It is presently called In-Situ Reactive Zones. I have asked Suthan Suthersan to assist in explaining this exciting new area.

The concept of In-Situ Reactive Zones is based on the creation of a subsurface zone, where migrating contaminants are intercepted and permanently immobilized or degraded into harmless end products. This is different from Reactive Walls and Funnel & Gate systems in which the groundwater flow pattern is also controlled. Reactive Zones allow groundwater to continue to flow naturally.

Successful design of Reactive Zones requires the ability to engineer the following types of in-situ reactions: (1) between the injected reagents and the subsurface environment in order to manipulate the biogeochemistry and optimize the required reactions, and (2) between the injected reagents and the migrating contaminants in order to effect remediation. These reactions will be different at each contaminated site and, in fact, may vary within a given site. Thus, the major challenge is to design a reactive zone(s) to systematically control these reactions under naturally variable or heterogeneous conditions found in the field.

The effectiveness of the Reactive Zone is determined largely by the relationship between the kinetics of the target reactions and the rate at which the mass flux of contaminants passes through the zone. Creation of a spatially fixed Reactive Zone in an aquifer requires not only the proper selection of reagents, but also the proper mixing of injected reagents uniformly within the Reactive Zone. Furthermore, such reagents must cause few side reactions and be relatively nontoxic in both original and treated forms.

The advantages of an In-Situ Reactive Zone to address the remediation of groundwater contamination are as follows:

- In-situ technology: eliminates the expensive infrastructure required for a pump and treat system; no disposal of water or wastes.
- Inexpensive installation: primary capital expenditure for this technology is the installation of injection wells.
- Inexpensive operation: reagents are injected at fairly low concentrations and, hence, the cost is insignificant; the only sampling required is groundwater quality monitoring.
- Can be used to remediate deep sites: no physical limits to design as with reactive walls.
- Unobtrusive: once the system is installed, site operations can continue without any obstructions.
- Immobilization of contaminants: utilizes the capacity of the soils and sediments to absorb, filter, and retain contaminants.

CONTAMINANT REMOVAL MECHANISMS

THE MECHANISMS USED TO REDUCE the toxicity of dissolved contaminants can be grouped into two major categories: transformation and immobilization. Conversion of chlorinated organic compounds to innocuous end products such as CO_2, H_2O, and Cl^- either by biotic or abiotic reaction pathways is an example of the transformation mechanism. Transformation of Cr(VI) to Cr(III) by either abiotic or biotic reaction pathways with precipitation as $Cr(OH)_3$ and subsequent removal by the soil matrix is an example of an immobilization mechanism.

Both of these mechanisms are going to be very important for remediation. However, due to space limitation in this column, we will have to limit our in-depth review to only one area. Since we normally do not have a chance to discuss metal treatment, we will concentrate on metal treatment in this article.

Immobilization mechanisms depend upon transforming the contaminant into a form that is much less soluble. Accordingly, transport of dissolved heavy metals in groundwater should be considered as a two-phase system in which the dissolved metals first transform and then partition between the soil matrix and the mobile aqueous phase.

Metal precipitates resulting from an in-situ reactive zone may move in association with colloidal particles or as particles themselves of colloidal dimensions (Vance, 1994). The term colloid is generally applied to particles with a size range of 0.001 to 1 micron. The transport of contaminants as colloids may result in unexpected mobility of low solubility precipitates. It is important to remember that the transport behavior of colloids is determined by the physical/chemical properties of the colloids as well as the soil matrix.

Colloidal precipitates larger than 2 microns in the low flow conditions common in aquifer systems will be removed by sedimentation (Vance, 1994). Colloidal precipitates are more often removed mechanically in the soil matrix. Mechanical removal of particles occurs most often by straining, a process in which particles can enter the matrix, but are caught by the smaller pore spaces as they traverse the matrix.

Colloidal particles below 0.1 micron will be subjected more to adsorptive mechanisms than mechanical processes. Adsorptive interactions of colloids may be affected by the ionic strength of the groundwater; ionic composition; quantity, nature, and size of the suspended colloids; geologic composition of the soil matrix; and flow velocity of the groundwater. Higher levels of total dissolved solids (TDS) in the groundwater encourage colloid deposition.

Chromium Removal

LET US FIRST LOOK AT an example of heavy metal removal. In-situ microbial reduction of dissolved hexavalent chromium Cr(VI) to trivalent chromium Cr(III) yields significant remedial benefits because trivalent chromium Cr(III) is less toxic, less mobile, and precipitates out of solution much more readily.

In-situ microbial reduction of Cr(VI) to Cr(III) can be promoted by injecting a carbohydrate solution, such as dilute molasses. The carbohydrates, which consist mostly of sucrose, are readily degraded by the heterotrophic microorganisms present in the aquifer, thus depleting all the available dissolved oxygen present in the groundwater. Depletion of the available oxygen present causes reducing conditions to develop. The mechanisms of Cr(VI) reduction to Cr(III), under the induced reducing conditions can be: (1) likely a microbial reduction process involving Cr(VI) as a terminal electron acceptor for the metabolism of carbohydrates, by species such as *Bacillus subtilis,* (Schroeder and Lee, 1975); (2) an extracellular reaction with by-products of sulfate reduction such as H_2S (DeFillippi, 1994); and (3) abiotic oxidation of the organic compounds including the soil organic matter such as humic and fulvic acids.

The primary end product of the Cr(VI) to Cr(III) reduction process is chromic hydroxide ($Cr[OH]_3$), which readily precipitates out of solution under alkaline to moderately acidic conditions. To ensure that this process will provide both short-term and long-term effectiveness in meeting groundwater cleanup objectives, the chromium precipitates must remain immobilized within the soil matrix of the aquifer and shall not be subject to dissolution or oxidation of Cr(III) back to Cr(VI) once groundwater reverts to its natural conditions. Based on the results of significant research being conducted on the in-situ chromium reduction process, it is readily apparent that the Cr(OH)$_3$ precipitate is essentially an insoluble, stable precipitate, immobilized in the soil matrix of the aquifer (Palmer and Puls, 1994).

Contrary to the numerous natural mechanisms that cause the reduction of Cr(VI) to Cr(III), there appear to be only a few natural mechanisms for the oxidation of Cr(III). Indeed, only two constituents in the subsurface environment (dissolved oxygen and manganese dioxide) are known to oxidize Cr(III) to Cr(VI) (Vance, 1994). The results of studies conducted on the potential reaction between dissolved oxygen and Cr(III) indicate that dissolved oxygen will not cause the oxidation of Cr(III) under normal groundwater conditions (Eary and Rai, 1987). However, studies have shown that Cr(III) can be oxidized by manganese dioxides which may be present in the soil matrix. However, only one phase of manganese dioxides is known to oxidize appreciable amounts of Cr(III) and this process is inversely related to groundwater pH. Hence, the oxidation of Cr(III) back to Cr(VI) in a natural aquifer system is highly unlikely.

The Cr(OH)$_3$ precipitate has an extremely low solubility (solubility product Ksp = 6.7×10^{-31}), and thus, very little of the chromium hydroxide is expected to remain in solution. It has been reported that aqueous concentration Cr(III), in equilibrium with Cr(OH)$_3$ precipitate, is less than 0.05 mg/L within the pH range of 5 to 12 (Palmer and Puls, 1994) (Figure 17.1). The pH range of natural aquifer systems will be within 5 to 12 and, hence, the potential for the chromic hydroxide to resolubilize is unlikely. Furthermore, the potential for coprecipitation with ferric ions will further decrease the solubility of Cr(OH)$_3$.

CASE HISTORY

A FIELD DEMONSTRATION STUDY WAS undertaken at an industrial facility in the midwest to demonstrate the efficacy of utilizing a microbially induced reduction process to convert hexavalent chromium to the less soluble (and less toxic) trivalent chromium. An existing 50 gpm pump and treat system was controlling the plume, but operation and maintenance was costing several hundred thousand dollars annually. The pilot study was undertaken at the source area, where hexavalent chromium concentrations were up to 15 ppm

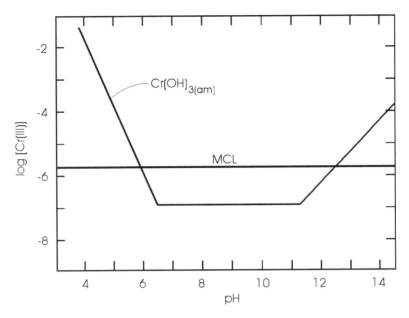

FIGURE 17.1. Cr(III) concentration in equilibrium with Cr(OH)$_3$.

(Figure 17.2). Three injection and five monitoring wells were installed and a carbon source consisting of a dilute molasses solution was periodically injected into the heart of the plume. As expected, the carbohydrate contained in the feed solution was readily degraded by the indigenous heterotrophic microorganisms present in the aquifer. The metabolic degradation process utilized the available dissolved oxygen contained in the groundwater, which caused strong reducing conditions to develop within the injection well zone within one month of process initiation. Under the developed reducing conditions, the hexavalent chromium Cr(VI) was reduced to trivalent chromium Cr(III). After three months of operation, wells that previously contained up to 15 ppm of Cr(VI) dropped to below 0.02 ppm, which was the groundwater cleanup objective (Figure 17.3). It is important to note that the groundwater cleanup objectives were met in unfiltered groundwater samples. This indicates that the chromium precipitate was being retained by the aquifer materials and was not subject to colloidal transport through the aquifer.

IN-SITU BIOTRANSFORMATION

THERE ARE TWO MAIN AREAS that have been studied for biotransformation: reductive dehalogenation and denitrification. We have discussed the treat-

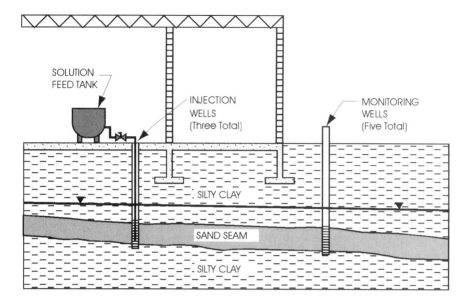

FIGURE 17.2. In-situ chromium reduction pilot study.

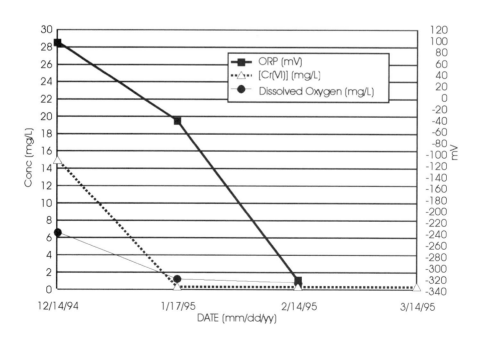

FIGURE 17.3. Graph of [Cr(VI)], DO and ORP levels in Injection Well 1 (IW-1).

ment of chlorinated hydrocarbons by reductive dehalogenation several times in this column before. The same sugar solutions that were used to create a reducing environment for Cr(VI) transformation can also create the reducing environment necessary for reductive dehalogenation. Other reactions are also possible for chlorinated hydrocarbons. In fact, this may be the most important application of Reactive Zones. However, I will save the subject of biotransformation of chlorinated hydrocarbons for a separate, future column when I can dedicate the appropriate space to fully explore the subject.

DENITRIFICATION

IN-SITU DENITRIFICATION CAN BE accomplished by organisms belonging to the genera *Micrococcus, Pseudomonas, Denitrobacillus, Spirillum, Bacillus,* and *Achromobacter,* which are usually present in the groundwater environment. Denitrifying organisms will utilize nitrate or nitrite in the absence of oxygen as the terminal electron acceptor for their metabolic activity. However, if any oxygen is present in the environment, it will probably be used preferentially. The energy for the denitrifying reactions is released by organic carbon sources which act as electron donors. The microbial pathways of denitrification include the reduction of nitrate to nitrite and the subsequent reduction of nitrite to nitrogen gas, as illustrated by the following equation:

$$NO_3^- \rightarrow NO_2^- \rightarrow N_2 \uparrow$$

In biological wastewater treatment processes employing denitrification, a cheap, external carbon source such as methanol is added as the electron donor. It has long been known that NO_3^- can be biodegraded to N_2 gas in anaerobic groundwater zones in the presence of a labile carbon source (Delwiche, 1981).

In-situ microbial denitrification is based on the same principle as conventional biological wastewater treatment systems, except that it is carried out in the subsurface by injecting the appropriate organic carbon source. Since methanol could be an objectionable substrate from a regulatory point of view, sucrose or sugar solution is a preferred substrate to be injected.

DESIGN OF IN-SITU REACTIVE ZONES

THE OPTIMIZATION OF SUBSURFACE ENVIRONMENTAL conditions to implement target reactions for remediating groundwater plumes holds a lot of promise. While some of the reactions may be relatively slow, long available residence times in the aquifer can be utilized as an advantage to implement cost-effective remediation strategies.

In-situ reactive zones can be designed as a curtain or multiple curtains to intercept the moving contaminant plume at various locations. An obvious choice for the location of an intercepting curtain is the downgradient edge of the plume. This curtain will act as a containment curtain to prevent further migration of the contaminants (Figure 17.4a). A curtain can be installed slightly downgradient of the source area to prevent the mass flux of contaminants migrating from the source (Figure 17.4b and c).

Another approach to designing an in-situ reactive zone is to create the reactive zone across the entire plume. The injection points can be designed on a grid pattern to achieve the reactions across the entire plume. However, it should be noted that the cost of installation of injection wells constitutes the biggest fraction of the system cost, looking at both capital and operational costs. Hence, it becomes very clear that the reduction of the total number of injection wells will significantly reduce the system costs.

Optimum Pore Water Chemistry

THE COMPOSITION OF INTERSTITIAL WATER is the most sensitive indicator of the types and the extent of reactions that will take place between contaminants and the injected reagents in the aqueous phase. Determination of the baseline conditions of the appropriate biogeochemical parameters is a key element for the design of an in-situ reactive zone. This evaluation will give a clear indication of the existing conditions and the necessary steps to be taken to optimize the environment to achieve the target reactions. A potential list of the biogeochemical parameters which should be determined for each site is presented below:

Dissolved oxygen
pH
Temperature
Redox potential
Total organic carbon (dissolved and total)
Total dissolved solids
Total suspended solids
NO_3^-
NO_2^-
SO_4^-
S^-
Fe (total and dissolved)

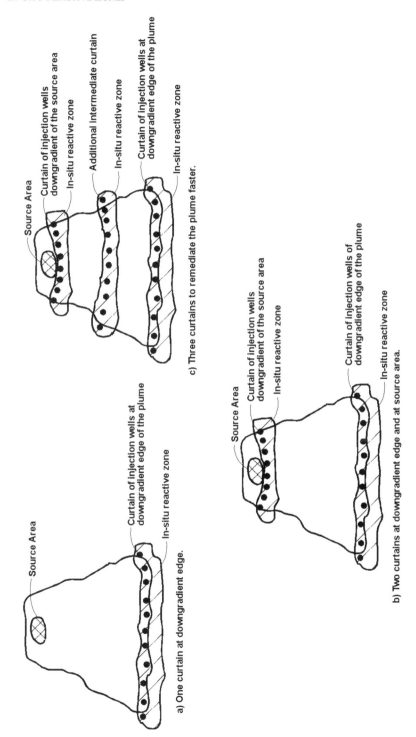

Source Area

Curtain of injection wells downgradient of the source area

In-situ reactive zone

Additional intermediate curtain

In-situ reactive zone

Curtain of injection wells at downgradient edge of the plume

In-situ reactive zone

c) Three curtains to remediate the plume faster.

Source Area

Curtain of injection wells at downgradient edge of the plume

In-situ reactive zone

a) One curtain at downgradient edge.

Source Area

Curtain of injection wells downgradient of the source area

In-situ reactive zone

Curtain of injection wells of downgradient edge of the plume

In-situ reactive zone

b) Two curtains at downgradient edge and at source area.

FIGURE 17.4. In-situ reactive zones based on curtain concept.

Mn (total and dissolved)

Carbonate content

Alkalinity

Concentration of dissolved gases (CO_2, N_2, CH_4, etc.)

Microbial population enumeration (total plate count and specific degraders count)

Any other organic or inorganic parameters that have the potential to interfere with the target reactions.

It should be noted that the number of parameters that need to be included in the list of baseline measurements will be site-specific and will be heavily influenced by the target reactions to be implemented within the reactive zone. The above list is a universal list and should be used solely as a reference list.

REGULATORY ISSUES

IN MOST CASES, IMPLEMENTATION OF an in-situ reactive zone requires injection of appropriate reagents and manipulation of the redox and biogeochemical environment within the reactive zone. Injection of reagents, albeit innocuous, nonhazardous and nonobjectionable, may raise some alarms regarding the short-term and long-term effects within the aquifer. Hence, considerable attention should be given for selecting an injection reagent which is easily acceptable and presents little or no side effects to the aquifer.

During immobilization reactions (for example, heavy metals precipitation), the contaminant is immobilized within the soil matrix below the water table. As noted earlier, under natural conditions this immobilization will be irreversible. Hence, the cleanup objective for the dissolved contaminant will be based upon the groundwater standard (for example, Cr[VI] = 10 µg/L), and when the contaminant is immobilized in the soil matrix, the cleanup standard will be based upon the soil standard (Cr[III] = 100 mg/kg). The huge difference in the two standards (for Cr, 10 ppb vs. 100 ppm) for the two phases is a significant benefit and provides a major advantage for achieving remediation objectives through an in-situ reactive zone (Puls, 1994).

In addition, consider a unit volume of the soil matrix below the water table, which has one liter of water in its pore spaces. Assuming a porosity of 30% and the specific gravity of the soil is 2.6, the same cube will have about 6.0 kg of soil. If the dissolved Cr(VI) concentration within the cube is 5 mg/L (ppm), the pore water within the cube contains 5 mg of Cr(VI) mass. When all this chromium is immobilized within the soil matrix of the cube, the concentration of the Cr(III) in the soil is equal to 0.83 mg/kg (ppm) (5 mg divided by 6.0 kg). Obviously, the small amount of Cr in the water has an insignificant effect on the concentration in the soil.

CONCLUSION

In-Situ Reactive Zones are one more bullet in the gun. These types of remediation will not be applicable to all sites, but they provide us with one more way to economically remediate contaminated sites. The reader should also remember that natural reactions encompass more than biological reactions. Abiotic reactions are an important part of understanding contamination plumes, their movement and remediation. I will revisit this technology from time to time as new applications are developed.

REFERENCES

1. DeFilippi, L. "Bioremediation of hexavalent chromium in water, soil and slag using sulfate reducing bacteria," In *Remediation of Hazardous Waste Contaminated Soils,* Wise, D. and D. Trantolo (Eds.), Marcel Dekker, New York, 1994.
2. Delwiche, C.C. "The nitrogen cycle and nitrous oxide." In *Denitrification, Nitrification and Atmospheric Nitrous Oxide,* Delwiche, C.C. (Ed.), Wiley-Interscience, New York, 1981.
3. Eary, L. and D. Rai. "Kinetics of chromium (III) oxidation to chromium (VI) by reaction with manganese dioxides," *Environ. Sci. Technol.,* 21(12):1187–1193, 1987.
4. Palmer, C. and R. Puls. Natural Attenuation of Hexavalent Chromium in Groundwater and Soils, USEPA, Office of Research and Development, Office of Solid Waste and Emergency Response, EPA/540/S-94/505, 1994.
5. Puls, R.W. Groundwater Sampling for Metals, Chapter 14, EPA Report No. EPA/600/A-94/172, 1994.
6. Schroeder, D. and G. Lee. "Potential transformations of chromium in natural waters," *J. Water Air, Soil Pollut.,* 4:355–365, 1975.
7. Vance, D.B. "Particulate transport in groundwater—I, Colloids," *Natl. Environ. J.,* Nov./Dec., 1994.

18

THE STATE-OF-THE-ART OF BIOREMEDIATION

Evan K. Nyer and Michael E. Duffin

HISTORY

IN THE BEGINNING THERE WAS Richard Raymond, and he gave his bacteria hydrogen peroxide and nutrients and told them to go forth and multiply. I know this because I was there (I'm old). I was also there when all of the companies that sold bacteria and nutrients swarmed into the market looking for large profits. And when a regional EPA office gave "official" approval to a bacteria that degraded PCBs. The company that had that bacteria made a nice living off doing research for the next several years, although no full-scale cleanups were completed. During those times, most of the regulators were in the "prove it" stage, and they wanted undeniable proof that the bacteria were responsible for all of the remediation that was going on. They were trying times. We knew bacteria were an important part of a remediation. However, due to the lack of scientific methods to prove biochemical reactions and the background noise caused by the specialized bacterial companies, most of the regulatory community would not accept in-situ biochemical designs, and called them a "do nothing" approach.

The technical division of EPA then took the "high road" and organized training seminars. Luckily, I was asked to join a ragtag group of biological experts as they traveled around the country trying to teach the basics of biology. While this started to provide a solid technical basis for the activity that was ongoing, most of us still just fumbled around, using bacteria the best that we could.

Finally, it all started to come together when the EPA led the effort to develop scientific methods of analysis as part of an in-situ biological design. This led to a real understanding of the biochemical reactions that were involved. More importantly, the regulators were finally given standard methods of testing and subsequent proof that natural attenuation was really working.

Of course, this is how I remember history. Many scientists, hydrogeologists, and engineers contributed to this long and fruitful effort. As they say, if you don't like my history, you are going to have to write your own article. What is important is that we now have the knowledge and the tools to apply biochemical reactions as part of our remedial designs. Let's look at the state of the art. I have asked Michael Duffin to help me summarize the published knowledge.

START WITH A DEFINITION

There have been many terms that have been used to describe the reactions that occur during a remediation. I think that "Natural Attenuation" does the best job of encompassing all of the processes that are occurring.

The U.S. EPA's Office of Research and Development and Office of Solid Waste and Energy Response defines the term "natural attenuation" as: *The biodegradation, dispersion, dilution, sorption, volatilization, and/or chemical and biochemical stabilization of contaminants to effectively reduce contaminant toxicity, mobility, or volume to levels that are protective of human health and the ecosystem.* As part of this definition, abiotic oxidation and hydrolysis are also attenuation mechanisms that destroy the contaminants to innocuous end products and thus reduce contaminant mass. However, biodegradation is the primary mechanism for attenuating biodegradable contaminants. (Wiedemeier et al., 1995; 1996). I will use the term bioremediation in this article to refer to the biochemical reactions of natural attenuation.

BIODEGRADATION PATHWAYS

The first thing that we have learned is the biodegradation pathways. From knowledge of these pathways we can measure the primary electron acceptors in the field. This information can be processed into rates of reaction and allows us to follow the biochemical reactions that are occurring during natural attenuation.

Mechanisms and Significance of Fuel-Hydrocarbon Biodegradation

As USED HERE, FUEL-HYDROCARBONS refer to the most commonly regulated compounds; i.e., benzene, toluene, ethylbenzene, and xylenes (BTEX). Almost all dissolved petroleum hydrocarbons are biodegradable under aerobic conditions, where microbes use oxygen as the electron acceptor and the contaminants (e.g., a benzene molecule), as the substrate for growth and energy. When oxygen is depleted and nitrate is present, anaerobic microbes will use nitrate as the electron acceptor. These first two reactions occur nearly instantaneously. Once the available oxygen and nitrate are depleted, the environment becomes more anaerobic and the slower reactions begin, such as microbes using ferric iron as an electron acceptor. When the reduction oxidation (redox) conditions are further reduced (usually near the source area, due to the abundant contaminant mass), sulfate may act as the electron acceptor. Under substantially lower redox conditions, methanogenic conditions will exist and the microbes can degrade the petroleum contaminants using water as the electron acceptor (Figure 18.1). Although the system above appears to be orderly with respect to species of electron acceptors, multiple reactions occur simultaneously in all biodegradation zones, including the continual reduction of dissolved oxygen (DO) owing to replenishment of all electron acceptors by natural groundwater recharge (Bouwer, 1994; Wiedemeier, et al., 1995).

The above system may be viewed as the assimilative capacity of the natural system in response to the introduction of contaminant mass flux across the aquifer. Thus, by knowing: (1) the volume of contaminated groundwater; (2) the background concentration of the specific electron acceptor (e.g., nitrate; ferric iron); and (3) the concentration of the specific electron acceptor measured in the contaminated area, it is possible to estimate the mass of BTEX lost to biodegradation through reduction of a specific electron acceptor. For example, each 1.0 mg/L of DO consumed by microbes destroys about 0.32 mg/L of BTEX; each 1.0 mg/L of ionic nitrate consumed by microbes destroys about 0.21 mg/L of BTEX; each 1.0 mg/L of sulfate consumed by microbes destroys about 0.21 mg/L of BTEX; and, conversely, degradation of 1.0 mg/L of BTEX results in the production of about 21.8 mg/L of iron II during iron III reduction and 0.78 mg/L of methane during methanogenesis (Wiedemeier, 1995).

In summary, the biodegradation of BTEX is mainly limited by electron acceptor availability, and biodegradation of these compounds will generally proceed until all the contaminants are destroyed. There appears to be an inexhaustible supply of electron acceptors in most, if not all, hydrogeologic environments. As evidence, in an estimated 80% of fuel hydrocarbon spills at

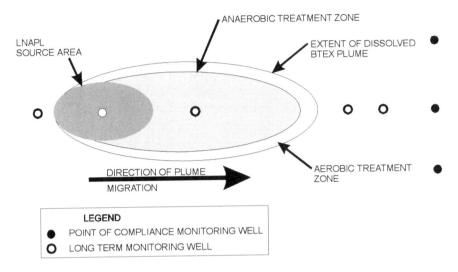

FIGURE 18.1. Biodegradation zones. From Wiedemeier, T.H., M.A. Swanson, D.E. Moutoux, J.T. Wilson, D.H. Kampbell, J.E. Hensen, and P. Hass, Overview of the Technical Protocol for Natural Attenuation of Chlorinated Aliphatic Hydrocarbons in Ground Water Under Development for the U.S. Air Force Center for Environmental Excellence. In: *Symposium on Natural Attenuation of Chlorinated Organics in Ground Water,* U.S. EPA 540/R-96/509, 1996.

federal facilities, natural attenuation alone, largely by bioremediation, will be protective of human health and the environment. Supporting data presented by Wiedemeier et al. (1996) for fuel spills at 40 air-force sites, and data presented by Rice et al. (1995), which included the study of 1,100 leaking underground storage tank sites in California, suggests that many fuel hydrocarbon plumes are relatively stable or are moving very slowly with respect to groundwater flow. Natural activity was preventing the spread of contamination. Based on the above trend in documented cases of bioremediation, state regulatory agencies are accepting bioremediation more often as a viable remedial alternative for contaminant containment and eventual cleanup at fuel-hydrocarbon spill sites.

Mechanism and Significance of Chlorinated Aliphatic Hydrocarbon Biodegradation

CHLORINATED HYDROCARBONS MAY UNDERGO BIODEGRADATION through three different pathways: (1) through use as an electron acceptor (reductive dechlorination); (2) through use as an electron donor; or (3) through co-metabolism.

Although all these processes may be occurring at a given site, the most important process for the natural biodegradation of the more highly chlorinated solvents, and that which will be discussed below, is reductive dechlorination. During reductive dechlorination, the chlorinated hydrocarbon is used as an electron acceptor, not as a source of carbon, and a chlorine atom is removed and replaced with a hydrogen atom. Reductive dechlorination generally occurs by sequential dechlorination from tetrachloroethene (PCE) to trichloroethene (TCE) to dichloroethene (DCE) to vinyl chloride (VC) to ethene (Figure 18.2). Thus, reductive dechlorination of chlorinated solvents is associated with an accumulation of daughter products and an increase in concentrations of chloride ions. Although the three isomers of DCE can theoretically be produced under reductive dechlorination, cis-1,2-DCE is a more common intermediate. If more than 80% of the total DCE is cis-1,2-DCE, the DCE is likely a reductive dechlorination-produced daughter product of TCE (Wiedemeier et al., 1996).

As with BTEX, the driving force behind oxidation-reduction reactions resulting in chlorinated hydrocarbons degradation is electron transfer, and the preferred sequence is similar to those reactions involving BTEX destruction. Both naturally occurring and anthropogenic organic carbon (fuel hydrocarbons) are used as electron donors. However, in general, biodegradation of fuel hydrocarbons is an electron-acceptor-limited process, while biodegradation of chlorinated hydrocarbons is an electron-donor-limited process.

DOCUMENTING BIOREMEDIATION

To support remediation by bacteria, the investigator must scientifically demonstrate that degradation of site contaminants is occurring at rates sufficient to be protective of human health and the environment, i.e., natural attenuation will reduce contaminant concentrations in groundwater to below risk-based corrective action or regulatory levels before potential receptor exposure pathways are completed. This is done by collecting and analyzing site-wide groundwater quality data (described in detail in Wiedemeier et al., 1995) with the following objectives:

- documented loss of contaminant mass at the field scale;
- biogeochemical indicator trends; and
- confirmation of microbial activity.

Documented loss of contaminant mass can be established by two methods. First, using chemical analytical data in mass balance calculations to show that decreases in contaminant and electron acceptor and donor concentrations

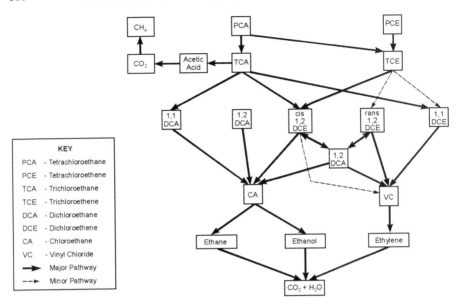

Figure 18.2. Transformations of chlorinated aliphatic hydrocarbons.

can be directly correlated to increases in metabolic end products or daughter compounds. This evidence can be used to show that electron acceptor and donor concentrations in groundwater are sufficient to facilitate degradation of dissolved contaminants. Second, using measured concentrations of contaminants and/or biologically recalcitrant tracers (e.g., trimethlylbenzene isomers for fuel-hydrocarbons and chloride for chlorinated hydrocarbons) in conjunction with aquifer hydrogeologic parameters, such as seepage velocity or dilution, to show that a reduction in contaminant mass is occurring at the site and to calculate biodegradation rate constants.

Biogeochemical trends can be established by collecting groundwater samples and analyzing for the following parameters: DO, redox potential, pH, temperature, electrical conductivity, alkalinity, nitrate, sulfate, ferrous iron, carbon dioxide, methane, chloride, and the contaminants of concern. The subsurface distribution of contamination and electron acceptor and metabolic by-product concentrations are important in documenting bioremediation. For example, if the DO concentration levels within the contamination plume are below background levels, it is an indication of aerobic biodegradation at those areas. Similarly, nitrate and sulfate concentrations below background levels in the plume are indications of anaerobic biodegradation through denitrification and sulfate reduction. Presence of nitrite and hydrogen sulfide in the plume further supports evidence of denitrification and sulfate reduction.

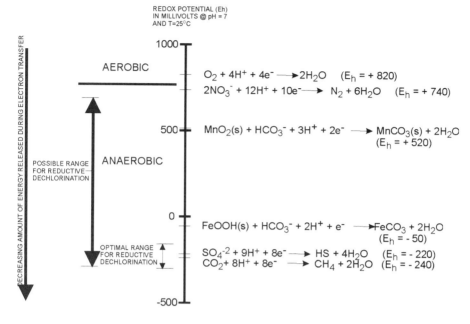

FIGURE 18.3. Redox potentials for various electron acceptors. (From Weidemeier, 1996.)

Elevated concentrations of metabolic by-products such as ferrous iron and methane will indicate the occurrence of Fe(III) reduction and methanogenesis inside the plume. Isopleth maps should be developed to show trends of these processes inside the contamination plume. Redox potential and hydrogen concentrations can be used to understand what types of reactions are occurring at the site. Figure 18.3 summarizes the redox potential and Table 18.1 summarizes the hydrogen concentration.

Microbiological activity can be confirmed through a microcosm study. This study involves using site aquifer materials under controlled conditions in the laboratory to show that indigenous microbes are capable of degrading site contaminants and to confirm rates of contaminant degradation measured at the field scale. Wilson et al. provides a good example of a microcosm study:

> In general, it is recommended that the investigation provide two of the three types of documentation. Not all of the data are available from all sites. Documenting the loss of mass may take several rounds of field samples. Microcosm studies may take up to 1 year for chlorinated hydrocarbons. Biogeochemical indications can be collected in one sampling event. Therefore, it is usually used at all sites with one of the other methods used to confirm the biochemical reactions.

TABLE 18.1. RANGE OF HYDROGEN CONCENTRATIONS FOR A GIVEN TERMINAL ELECTRON-ACCEPTING PROCESS[a]

Terminal Electron-Accepting Process	Hydrogen Concentration (nanomoles per liter)
Denitrification	< 0.1
Iron(III) reduction	0.2 to 0.8
Sulfate reduction	1 to 4
Methanogenesis	> 5

From Wiedemeier, T.H., M.A. Swanson, D.E. Moutoux, J.T. Wilson, D.H. Kampbell, J.E. Hensen, and P. Hass, Overview of the Technical Protocol for Natural Attenuation of Chlorinated Aliphatic Hydrocarbons in Ground Water Under Development for the U.S. Air Force Center for Environmental Excellence. In: *Symposium on Natural Attenuation of Chlorinated Organics in Ground Water*, U.S. EPA 540/R-96/509, 1996.

DOES IT WORK?

BIOREMEDIATION OF HYDROCARBONS OCCURS IN three plume zones: (1) the source area zone—core of the plume; (2) the anaerobic treatment zone; and (3) the aerobic treatment zone (Figure 18.1; Wiedemeier et al., 1995; 1996). The changes in the biogeochemical parameters can be used to calculate the natural capacity of the aquifer across these zones. Since we are measuring direct by-products of the reactions that are occurring, the calculated capacity can be fairly accurate. This confirmation can be summarized in an analytical model. BIOSCREEN is a nonproprietary model that is available from the Robert S. Kerr Laboratory's home page on the Internet (www.epa.gov/ada/kerrlab.html). I suggest that you all download this very simple-to-use model and try it.

Chlorinated plumes are more difficult to quantify. Contaminant plumes formed by chlorinated hydrocarbons dissolved in groundwater can show three types of behavior (Type 1, 2, and 3), based on the amount and type of primary substrate present in the aquifer materials. Type 1 plumes occur where anthropogenic carbon (e.g., BTEX from a fuel spill), is being utilized as the primary substrate for microbial degradation. Such plumes are typically anaerobic and conditions result in rapid and extensive degradation, via reductive dechlorination, of the highly chlorinated hydrocarbons (e.g., PCE, TCE and DCE). For example, using the equation given in Wiedemeier et al., 1996, in a Type 1 plume, oxidation of BTEX via reductive dechlorination would consume 12.6 mg of TCE. Type 2 plumes occur in areas that are characterized by high natural organic carbon concentrations and anaerobic conditions. Microbes utilize the natural organic carbon as a primary substrate; if redox conditions are favorable (i.e., areas with high natural organic carbon contents), highly chlorinated hydrocarbons can be rapidly degraded via reductive dechlorination. For example, using the equation given in Wiedemeier et al., 1996, oxidation of natural, dissolved organic carbon via reductive dechlorination would consume 10.5 mg TCE. Type 3 plumes occur in areas characterized by low concentrations of natural and/or anthropogenic carbon, and by DO concentrations exceeding 1.0 mg/L. Biodegradation of chlorinated hydrocarbons via reductive dechlorination will not occur under these conditions; thus, there is no mass removal of PCE, TCE or DCE, only attenuation via dilution (e.g., by advection, dispersion, sorption). However, biodegradation of less chlorinated compounds such as VC can occur via oxidation. A single plume, however, can show all three of the plume-type characteristics described above. An example of the most beneficial scenario is a mixed-type plume where PCE, TCE, and DCE undergo reductive dechlorination (Type 1 and 2 plumes), then VC is oxidized (Type 3 plume).

Most sites have indigenous bacteria capable of promoting chlorinated hydrocarbons biodegradation. In a study of over 50 DuPont sites with chlorinated hydrocarbons plumes, Ellis (1996) documented that 88% of the sites have bacteria that can biodegrade PCE and TCE to DCE, 75% of the sites have bacteria that can biodegrade DCE to VC or ethene, and 58% of the sites have bacteria that can biodegrade VC to ethene. However, as Cherry (1996) noted, unless the biodegradation is sufficiently complete, biodegradation may render the plume more toxic. Even so, under most situations, bioremediation is clearly the method of remediation to consider when compared to the conventional pump and treat method.

To determine if bioremediation of chlorinated hydrocarbons is occurring, this protocol suggests sampling at least six wells that appear to be representative of the contaminant flow system and analyzing the parameters listed in Table 18.2. Samples should be collected from: (1) the most contaminated portion of the aquifer; e.g., spill zone; (2) downgradient from the NAPL source area but still in the dissolved contaminant plume; (3) downgradient from the dissolved contaminant plume; and (4) from both upgradient and lateral locations that are not affected by the plume. After these samples have been analyzed for the parameters, the investigator should analyze the data to determine whether biodegradation is occurring. The right-hand column of Table 18.2 contains scoring values that can be used for this evaluation. Table 18.3 summarizes the range of possible scores and gives an interpretation for each score. If the site scores a total of 15 or more points, biodegradation is probably occurring, and the investigator should proceed to step 2. However, if the site scores below 15 it does not necessarily mean biodegradation is not occurring.

TABLE 18.2. ANALYTICAL PARAMETERS AND WEIGHTING FOR PRELIMINARY SCREENING[a]

Analyte	Concentration in Most Contaminated Zone	Interpretation	Points Award
Oxygen[b]	<0.5 mg/L	Tolerated; suppresses reductive dechlorination at higher concentrations	3
Oxygen[b]	>1 mg/L	Vinyl chloride may be oxidized aerobically, but reductive dechlorination will not occur	−3
Nitrate[b]	<1 mg/L	May compete with reductive pathway at higher concentrations	2
Iron (II)[b]	>1 mg/L	Reductive pathway possible	3
Sulfate[b]	<20 mg/L	May compete with reductive pathway at higher concentrations	2
Sulfide[b]	>1 mg/L	Reductive pathway possible	3
Methane[b]	>0.1 mg/L	Ultimate reductive daughter product	2
	>1	Vinyl chloride accumulates	3
	<1	Vinyl chloride oxidizes	
Oxidation reduction potential[b]	<50 mV against Ag/AgCl	Reductive pathway possible	<50 mV = 1 <−100 mV=2
pH[b]	5<pH<9	Tolerated range for reductive pathway	
DOC	>20 mg/L	Carbon and energy source; drives dechlorination; can be natural or anthropogenic	2
Temperature[b]	>20°C	At T>20EC, biochemical process is accelerated	1
Carbon dioxide	>2x background	Ultimate oxidative daughter product	1
Alkalinity	>2x background	Results from interaction of carbon dioxide with aquifer minerals	1
Chloride[b]	>2x background	Daughter product of organic chlorine; compare chloride in plume to background conditions	2

Analysis	Concentration	Interpretation	Points
Hydrogen	>1 nM	Reductive pathway possible; vinyl chloride may accumulate	3
Hydrogen	<1 nM	Vinyl chloride oxidized	
Volatile fatty acids	>0.1 mg/L	Intermediates resulting from biodegradation of aromatic compounds; carbon and energy source	2
BTEX[b]	>0.1 mg/L	Carbon and energy source; drives dechlorination	2
Perchloroethene[b]		Material released	
Trichloroethene[b]		Material released or daughter product of perchloroethene	2[c]
Dichloroethene[b]		Material released or daughter product of trichloroethene; if amount of cis-1,2-dichloroethene is greater than 80% of total dichloroethene, it is likely a daughter product of trichloroethene	2[c]
Vinyl chloride[b]		Material released or daughter product of dichloroethenes	2[c]
Ethene/Ethane	<0.1 mg/L	Daughter product of vinyl chloride/ethene	>0.01 mg/L=2; >0.1 = 3
Chloroethane[b]		Daughter product of vinyl chloride under reducing conditions	2
1,1,1-Trichloroethane[b]		Material released	
1,1-Dichloroethene[b]		Daughter product of trichloroethene or chemical reaction of 1,1,1-trichloroethane	

[a] From Wiedemeier, T.H., M.A. Swanson, D.E. Moutoux, J.T. Wilson, D.H. Kampbell, J.E. Hensen, and P. Hass, Overview of the Technical Protocol for Natural Attenuation of Chlorinated Aliphatic Hydrocarbons in Ground Water Under Development for the U.S. Air Force Center for Environmental Excellence. In: *Symposium on Natural Attenuation of Chlorinated Organics in Ground Water*, U.S. EPA 540/R-96/509, 1996.
[b] Required analysis.
[c] Points awarded only if it can be shown that the compound is a daughter product (i.e., not a constituent of the source NAPL).

TABLE 18.3. INTERPRETATION OF POINTS AWARDED DURING SCREENING STEP 1[a]

Score	Interpretation
0 to 5	Inadequate evidence for biodegradation of chlorinated organics
6 to 14	Limited evidence for biodegradation of chlorinated organics
15 to 20	Adequate evidence for biodegradation of chlorinated organics
>20	Strong evidence for biodegradation of chlorinated organics

[a] From Wiedemeier, T.H., M.A. Swanson, D.E. Moutoux, J.T. Wilson, D.H. Kampbell, J.E. Hensen, and P. Hass, Overview of the Technical Protocol for Natural Attenuation of Chlorinated Aliphatic Hydrocarbons in Ground Water Under Development for the U.S. Air Force Center for Environmental Excellence. In: *Symposium on Natural Attenuation of Chlorinated Organics in Ground Water,* U.S. EPA 540/R-96/509, 1996.

For example, if a site scores low, say 7 points, but has a large amount of cis-1,2-DCE in the system, then this is conclusive evidence that biodegradation of TCE is occurring (Ellis, 1996).

CONCLUSION

WE HAVE COME A LONG way. Due to the hard work that has been done in the field during the past several years, we are now able to make bacteria part of our remedial designs. I will let John Wilson (1996) have the last words:

- Bioremediation is real and can be protective when properly implemented;
- Bioremediation occurs at many sites;
- Each biodegradation step has an average half-life of 1 to 2 years;
- The most significant factors in determining the effectiveness of bioremediation are plume residence time and the half-lives of the sequential biodegradation reactions;
- Most chlorinated hydrocarbons plumes eventually will reach steady-state equilibrium and no longer expand;
- Most aquifer sediments contain much more natural organic carbon than necessary to support bioremediation; and
- Bioremediation is not a "do nothing" approach, and there is considerable cost associated with the effort. However, the total costs for the bioremediation alternative are about 60% less than for the cheapest pump and treat system.

REFERENCES

Bouwer, E.J. Bioremediation of Chlorinated Solvents Using Alternate Electron Acceptors. In: *Handbook of Bioremediation,* Norris, R.D., R.E. Hinchee, R.

Brown, P.L. McCarty, L. Semprini, J.T. Wilson, D.H. Kampbell, M. Reinhard, E.J. Bouwer, R.C. Borden, T.M. Vogel, J.M. Thomas, and C.H. Ward (Eds.), Lewis Publishers, Boca Raton, FL, 1994.

Cherry, J.A. Conceptual Models for Chlorinated Solvent Plumes and their Relevance to Intrinsic Remediation. In: *Symposium on Natural Attenuation of Chlorinated Organics in Ground Water*, U.S. EPA, 540/R-96/509, 1996, pp. 29–30.

Ellis, D.E., Intrinsic Remediation in the Marketplace. In: *Symposium on Natural Attenuation of Chlorinated Organics in Ground Water*, U.S. EPA, 540/R-96/509, 1996, pp. 120–123.

Nyer, E.K., P.L. Palmer, T.L. Crossman, G. Boettcher, S.S. Suthersan, S. Fam, and D.F. Kidd. *In-Situ Treatment Technology*, CRC Press/Lewis Publishers, Boca Raton, FL, 1996.

Wiedemeier, T.H., J.T. Wilson, D.H. Kampbell, R.N. Miller, and J.E. Hansen. Technical Protocol for Implementing Intrinsic Remediation with Long-term Monitoring for Natural Attenuation of Fuel Contamination Dissolved in Groundwater, U.S. Air Force Center for Environmental Excellence, San Antonio, TX, 1995.

Wiedemeier, T.H., M.A. Swanson, D.E. Moutoux, J.T. Wilson, D.H. Kampbell, J.E. Hansen, and P. Hass. Overview of the Technical Protocol for Natural Attenuation of Chlorinated Aliphatic Hydrocarbons in Ground Water Under Development for the U.S. Air Force Center for Environmental Excellence, In: *Symposium on Natural Attenuation of Chlorinated Organics in Ground Water*, U.S. EPA 540/R-96\509, 1996, pp. 35–59.

Wilson, B.H., J.T. Wilson, and D. Luce. *Design and Interpretation of Microcosm Studies for Chlorinated Compounds*, U.S. EPA, pp. 21–27.

Wilson, J.T., D.H. Kampbell, and J.W. Weaver. Environmental Chemistry and the Kinetics of Biotransformation of Chlorinated Organic Compounds in Ground Water, in: *Symposium on Natural Attenuation of Chlorinated Organics in Ground Water*, U.S. EPA 540/R-96/509, 1996, pp. 124–127.

PART V

ADVANCED REMEDIATION TECHNIQUES

19

AIR SPARGING–SAVIOR OF GROUNDWATER REMEDIATIONS OR JUST BLOWING BUBBLES IN THE BATHTUB

Evan K. Nyer and Suthan S. Suthersan

INTRODUCTION

AIR SPARGING IS HOT. EVERYONE is using this technology. And, in the great environmental industry tradition, people have decided that sparging can treat anything under any condition. I don't know what it is about this industry, but we seem to constantly take good technology and overapply it. Any technology applied in the wrong circumstances can fail. We then have a good technology with a bad reputation.

I have asked Suthan Suthersan of Geraghty & Miller's Plainview office to assist me in this article. Suthan is responsible for application of new technology for Geraghty & Miller. During the past couple of years, he has installed several sparging systems. We have written this article to be an introduction to air sparging. Hopefully, it will help limit the applications of this important technology, and preserve its reputation.

In-situ air sparging has only recently been employed as a technology for the remediation of dissolved volatile organic compounds (VOCs). The process could be aptly called "in-situ air stripping" due to the primary mode of mass transfer during its operation. In-situ air sparging is often used in conjunction with vacuum extraction systems to remove the stripped contaminants, and has broad appeal due to its simplicity and low costs. The difficulties encountered in modeling the multiphase air sparging process (i.e., air flow in saturated conditions) have made this technology dependent on empirical information for the engineering design of these systems. At this point, it is as important to know "what not to do" as "what to do" when designing an in-situ air sparging system.

APPLICABILITY AND LIMITATIONS

IN-SITU AIR SPARGING BECOMES a very attractive remedial option when volatile and/or easily biodegradable organic contaminants are present in the groundwater. The in-situ air sparging process can be defined as injection of compressed air at controlled pressures and volumes in the groundwater table below the deepest known point of the contamination. Figure 19.1 represents a typical air sparging process schematic.

There are three mass transfer phenomenons which take place during this process: in-situ stripping of dissolved VOCs, volatilization of adsorbed phased contamination below the water table, and enhanced biodegradation of both dissolved and adsorbed phase contaminants due to the increase in dissolved oxygen levels.

Among the above mass transfer processes, the contaminant applicability will be mainly influenced by the strippability of the VOCs. The strippability of any contaminant is influenced by its Henry's law constant, and compounds such as benzene, toluene, xylene, ethylbenzene, trichloroethylene, and tetrachloroethylene are considered to be very easily strippable. (See Nyer, 1992 for the Henry's law constant for 50 compounds). Direct volatilization of a compound is governed by its vapor pressure and most volatile organic compounds are easily volatilized by their definition. Biodegradability of any compound under aerobic conditions is dependent on its chemical structure and the environmental parameters. Some of the VOCs are considered to be easily biodegradable under aerobic conditions (e.g., petroleum compounds, acetone, etc.) and some of them are not (e.g., chlorinated VOCs). Table 19.1 describes the contaminant applicability of various VOCs for in-situ air sparging.

This is the good news. The bad news is that there are limitations beyond the properties of the compounds themselves. The geology of the site must also be considered when applying air sparging. Physical implementation of in-situ air sparging is greatly influenced by the ability of the injected air bubbles to travel in a vertical direction. The two main geology limitations are air (or hydraulic) conductivity, and homogeneity.

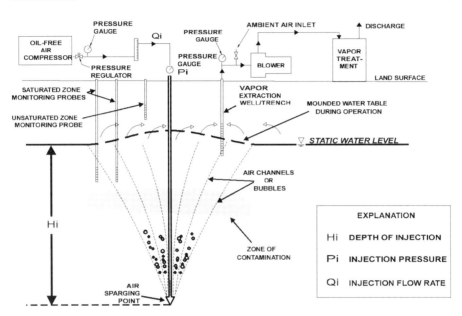

FIGURE 19.1. Air sparging process schematic.

TABLE 19.1. CONTAMINANT APPLICABILITY FOR IN-SITU AIR SPARGING

Contaminant	Strippability	Volatility	Biodegradability
TCE	High	High	Very low
PCE	High	High	Very low
Acetone	Very low	High	High
Benzene	High	High	High
Toluene	High	High	High
Xylenes	High	High	High
MTBE	Moderate	Moderate	Moderate
Gasoline constituents	High	High	High
Fuel oil constituents	Low	Low	Moderate

As with any technology that relies on a fluid moving through the aquifer, air sparging requires certain conductivity in order for the air to move in an appropriate manner. Problems can occur if the permeability is too high or too low. If the pores are too large, e.g., gravel, then the air has no horizontal movement. Air released at the well only travels vertically, and very little area is covered by the sparging system. The next section reviews the "cone of influence" from a sparging well.

If the soil is too tight, e.g., clay, then two problems can occur. The soil can provide too much resistance to air travel through the system, and/or the vertical movement can be too large, and the sparging system can actually spread the contaminants. It is important to remember that we are discussing the vertical permeability of the soil and not the horizontal permeability when we discuss air sparging. The vertical permeability can be 10 to 1,000 times less permeable than the horizontal.

The other main geological problem is when the aquifer is not homogeneous. Under these conditions, the air can travel vertically until a tight layer of soil is encountered, and then the air may travel horizontally. Once again, this may actually spread the contaminants.

Other problems that can occur are as follows:

- The injected air displaces the pore water during its passage toward the vadose zone. Due to this displacement, mounding of water would occur at the top of the cone of influence. The height of mounding is dependent on the volume of water displaced and thus is directly proportional to the amount of air injected. It becomes very apparent that this mounding may cause problems in controlling the movement and capture of any free product layer floating on top of the water table. If the free product thickness is in the order of only a sheen, the mass volatilization taking place at the surface due to air sparging will minimize the negative impact.
- Continuous injection of air into the water table and the resulting displacement of pore water may lead to formation of air pockets and thus cause resistance to the movement of the groundwater.
- The structural stability of foundations or utilities can be in jeopardy.
- There is some risk of uncontrolled contaminant vapor migration to basements or ignition sources.

SYSTEM DESIGN PARAMETERS

IN THE ABSENCE OF ANY reliable models to define the in-situ air sparging process, empirical approaches are the most reliable means of designing a system. The main parameters that are of significant importance in designing an in-situ air sparging system are listed below.

- radius of the "cone of influence."
- depth of air injection.
- air injection pressure.

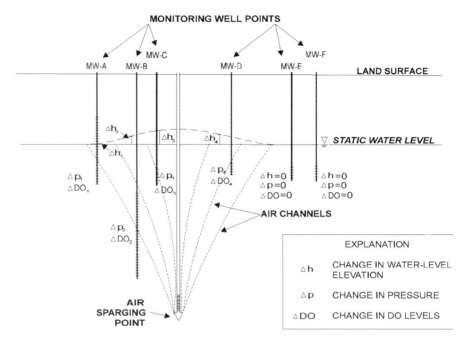

FIGURE 19.2. Air sparging test measurements.

Radius of Influence

THE RADIUS OF THE "CONE of influence" can only be measured by performing a field design test. The dimensions of the cone (Figure 19.2) depend on the angle of distribution of air and the depth of injection. The angle of distribution, in turn, seems to be dependent on the inherent geologic characteristics, and injection air flow rate. The typical range of the angle of distribution seems to be between 15° (coarse gravels) to 60° (silty sands). Any attempt to increase the angle beyond 60° by injecting large volumes of air will lead to "pushing" the contaminants in the horizontal direction.

Currently, there is no standard method for determining the radius of influence during the field testing of an in-situ air sparging system. Potential measuring parameters (Figure 19.2) include: measurement of the lateral extent of groundwater mounding, increase in dissolved oxygen levels, and redox potential in comparison to presparging conditions; increase in positive pressure within the cone; or the actual reduction in VOC concentrations due to sparging.

Among the above parameters, positive pressure distribution seems to be the most reliable measurement since it directly represents the forced displacement of porewater. Using the lateral extent of groundwater mounding may slightly overestimate the actual radius of influence; since it is hard to define the portion of the mound caused by the displacement of pore water at the surface versus the total displacement within the cone of influence. Measurement of dissolved oxygen levels may also overestimate the radius of influence due to the potential for diffusional transport within the time required for equilibration of the cone formation.

Depth of Air Injection

AMONG ALL THE DESIGN PARAMETERS, this may be the easiest to determine, since the choice is very much influenced by the contaminant distribution. It is prudent to choose the depth of injection at least a foot or two deeper than the "deepest known point" of contamination. It should be noted that *sparging depth* determination should be influenced only by the depth below the water table.

There are no known absolute limitations in choosing the sparging depth. The current experience in the industry is based on depths less than 30 to 40 feet. At depths greater than 40 feet, due to the need for a large volume of air, nested injection points are recommended. This may overcome the potential for channeling, due to the presence of micropockets of higher permeability across larger depths. The apparent reluctance to try sparging at deeper depth is also because of the fear of safety considerations due to the high pressures involved.

Air Injection Pressure

THE INJECTION PRESSURE NECESSARY TO initiate in-situ air sparging should be able to overcome the following:

1. the depth of water column at the point of injection.
2. the frictional losses in the pipes in the well.
3. the capillary entry resistance to displace the pore water; this depends on the type of sediments in the subsurface.

Hence, the Pressure of Injection (P_i) in feet of water could be defined as:

$$P_i = H_i + P_R$$
H_i = depth of injection (feet of water).
P_R = release pressure, which is a combination of (2) and (3) above (feet of water).

In addition, the following parameters also need to be reviewed for air sparging design.

- air flow rate
- injection well design
- air distribution efficiency.

The release pressure (P_R) depends on the type of geology, depth of injection, air distribution efficiency, and the frictional losses in the system. Release pressure has been found to be in the range of 2.3 feet of H_2O (1 psi) for every 3 feet of H_i for fine sands and every 5 feet of H_i for coarse gravel.

The notion that the effect of air sparging will be better, higher the pressure, is not true. In contrast, the negative impacts due to higher pressures than required will be manifested in the form of uncontrolled migration of contaminants due to increased turbulence. Also, higher pressures may increase the operating cost of the system, as described in the next section.

Injection Wells

THE AMOUNT OF AIR FLOWING in an air sparging well is relatively small. Hence, the "ideal" diameter for air sparging wells is the smallest possible well which could be installed to the required depth. Based on these requirements, the authors have found driven well points made out of 8 to 10 feet cast iron-flush jointed sections (Figure 19.1) to be extremely useful to install air sparging points in the 3/4-inch to 1/2-inch diameter range.

Injection Air Flow Rate

INJECTED AIR DISPLACES THE PORE water during its passage toward the vadose zone. Hence, the air thus replacing the pore water will tend to divert a portion of the groundwater flow around the cone of influence. This could be easily overcome by pulsing the injection, and by injecting the minimum amount of air to achieve the maximum dimensions of the cone of influence. As described earlier, the maximum attainable angle of the cone during equilibrium is an inherent characteristic of the geology. The authors have found that the maximum dimension of the cone could be achieved by injecting as little as 30% of the pore volume for finer sediments and 50% for coarser sediments.

Typical air sparging flow rates lie in the range of 4 cfm (for depths less than 10 feet) to 10 cfm up to depths 30 feet. Very low flow rates may not be sufficient for the air bubbles containing the contaminants to reach the vadose zone, and very high flow rates may again produce turbulent flow and may contribute toward uncontrolled migration of the contaminants.

Air Distribution Efficiency

THE STANDARD AIR SPARGING WELL design consisted of a sparging point made up of screens of length (6 inches to 2 feet), with bentonite seal above it in the annular space in the borehole to prevent short-circuiting of the injected air. Since the authors encourage the use of small diameter driven well points, screens or very small diameter holes can be used as the exit points for the air. The authors' experience indicates that the use of very small diameter holes (1/16 inch to 3/16 inch) has an impact on increasing the radius of influence under similar flow rates and depths. This could be due to a higher horizontal gradient of the exit velocity when the air exits through the available area in the holes in comparison to the screen.

SUCCESSES AND FAILURES

AS CAN BE SEEN BY the previous sections, air sparging has tremendous potential if it is applied under the right conditions. The problem is that even when the application is correct, air sparging is still not a "cure-all." The technology can do many things, but not everything.

Sparging is great in removing large amounts of contaminants in a relatively short period of time.

CLEANUP RATES

CLEANUP TIMES OF LESS THAN 12 months have been reported in the literature. The authors have seen a reduction of cyclohexane from about 40 mg/L in the groundwater to less than 5 mg/L in a period of 8 months. They have also seen TCE reduction from 1000 μg/L to 400 μg/L in six months after the site had already been treated with pump and treat and vapor extraction for several years. The actual cleanup time will depend on the phase distribution of contaminant(s) below the water table (i.e., the fraction in the dissolved phase versus the adsorbed phase), the strippability, biodegradability of the contaminants, density of sparging points, hydrogeologic characteristics, etc.

The problem has been that sparging systems seem to also reach an asymptote, just as groundwater pump and treat systems do. While the first part of the Life Cycle curve is steeper, and the level of the asymptote is lower, the system still reaches an asymptote and further reduction of the concentration is very slow.

This makes sense when compared to recent research on air sparging. It seems that when the air is released from the well or well point, it does not rise as separate, discrete bubbles. The rising air creates channels in the aquifer, and all of the air rises through these channels. These channels are close enough together to directly contact significant portions of the aquifer. However, remediation of the area in between the channels is still controlled by diffusion from those areas to the channels. This creates the asymptote and can finally limit the concentration that is reached in the groundwater.

REFERENCES

Nyer, E.K. *Practical Techniques for Groundwater and Soil Remediation,* Lewis Publishers, Boca Raton, FL, 1993.
Johnson, R.C. Personal communication, 1993.

FURTHER READING

Sellers, K.L. and R.P. Schreiber, "Air Sparging Model for Predicting Groundwater Cleanup Rates," Proceedings of the Petroleum Hydrocarbons and Organic Chemicals in Groundwater: Prevention, Detection and Restoration, Houston, 1992.
Arendt, F., M. Hinsenveld, and W.J. Van den Brink (Eds.). "Air injection and soil air extraction as a combined method for cleaning contaminated sites: Observations from test sites in sediments and solid rocks." In: *Contaminated Soil,* Kluwer Academic Publishers, 1990, pp. 1039–1044.
Marley, M.C., F. Li, and S. Magee. "The Application of a 3-D Model in the Design of Air Sparging Systems," Proceedings of the Petroleum Hydrocarbons and Organic Chemicals in Groundwater: Prevention, Detection and Restoration, Houston, TX, 1992.
Brown, R.H., C. Herman, and G. Henry. The Use of Aeration in Environmental Clean Ups, Proceedings of Haztech International Pittsburgh Waste Conference, 1991.

20

THERE ARE NO IN-SITU METHODS

Evan K. Nyer and David C. Schafer

WE WILL NEVER UNDERSTAND THE strengths and limitations of the new innovative technologies unless we comprehend how they accomplish our remediation goals. Too many of our sites will take 10 to 100 years, or more, to reach final cleanup goals. We do not want to wait until the end of the project to determine if the technology will accomplish full cleanup.

The first step to reach this understanding is to realize how the technology works. The two hottest "in-situ" technologies being applied in the field today are Soil Vapor Extraction (SVE; also known as Vapor Extraction Systems, VES), and Air Sparging. Neither of these techniques are, in fact, in-situ methods. Both of these technologies rely on air movement to remove the contaminants from the ground and aquifer. This does not constitute an "in-place" treatment; it is a simple change of carrier. Pump and treat techniques use water as a carrier to remove the contaminants from the aquifer. SVE and Air Sparging use air as the carrier to remove contaminants from the ground and aquifer. Air provides several advantages over water, but still has some of the weaknesses of water.

WATER VS. AIR AS THE CARRIER

ONE OF THE IMPORTANT ADVANTAGES in switching from water to air as the carrier is the number of pore volumes that can be processed through the soil or aquifer in a short period of time.

Water as the Carrier

AN ELLIPTICAL CONTAMINATION PLUME 80 ft wide by 300 ft long must be hydraulically contained in an aquifer 40 ft thick having a hydraulic conductivity of 30 ft/day. The natural gradient at the site is measured at 0.01 ft/ft. Thus, the relevant parameters are as follows:

K = hydraulic conductivity = 30 ft/day
b = aquifer thickness = 40 ft
I = hydraulic gradient = 0.01
T = K × b = transmissivity = 1,200 ft²/day

To hydraulically contain the plume, the flow rate may be computed as follows:

$$Q = 2TIW_0$$

where Q = flow rate, in cfd
T = transmissivity, in ft²/day
I = hydraulic gradient, in ft/ft
W_0 = capture width at the well, in ft

From this,

$$Q = 2 \bullet 1200 \bullet 0.01 \bullet 80$$
$$= 1,920 \text{ cfd or 10 gpm}$$

Figure 20.1 shows the plume and capture zone boundary for a 10 gpm well located 25 ft from the leading edge of the plume.

Analytic element modeling was used to determine the flushing rate by computing the travel time from the upgradient edge of the plume to the recovery well. Figure 20.2 shows a streamline approaching the recovery well, and includes tick marks at 10-day intervals. The total travel time from the upgradient edge of the plume to the recovery well is 175 days based on an assumed porosity of 25%. Thus, a conservative estimate of pore volume exchange rate is 1 every 175 days.

During this time, however, clean water outside the plume, but within the capture zone, converges through the plume to the recovery well. This has the effect of increasing the number of pore volume flushings. To account for this, another way to estimate the number of pore volume exchanges is to simply compare the volume of water in the plume to the extraction rate of 10 gpm.

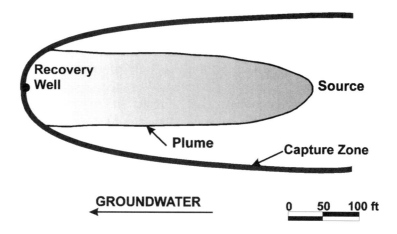

FIGURE 20.1. Capture zone for 10-gpm pumping rate.

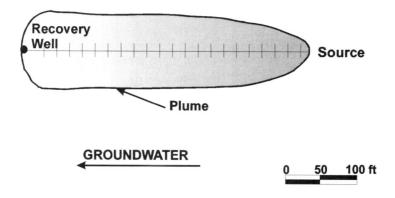

FIGURE 20.2. Ten day time intervals along streamline.

Approximating the plume as an 80-ft by 300-ft ellipse, the volume of water it contains is expressed as follows:

$$V = \pi \bullet \frac{300}{2} \bullet \frac{80}{2} \bullet 40 \bullet 0.25$$
$$= 188,495 \text{ ft}^3$$

Dividing by the flow rate of 1,920 cfd (10 gpm) gives an average flushing time of 98 days. We will use this number is comparing pore volume exchanges.

Air as the Carrier

ASSUME NOW THAT WE HAVE a similar area of soil contamination above the water table that will be cleaned up using vapor extraction wells. Assume further that the thickness of the vadose zone is 40 ft, the same as the assumed aquifer thickness in the previous example. Figure 20.3 shows a typical design we might use to vapor extract contaminants using four wells running along the axis of the contaminated area. Often we select flow rates for the vapor extraction wells to produce from 1 to 4 pore volume exchanges per day. For this example, we will aim for 2 pore volume exchanges per day.

Using the same porosity as before, 25%, the air volume within the contaminated zone is 188,495 cubic ft. The total air flow rate required for 2 pore volume exchanges per day is

$$Q = \frac{2 \bullet 188,495}{1,440}$$
$$= 262 \text{ cfm}$$

This requires an average flow rate of 65.5 cfm per well. To see if this is a reasonable expectation, we will use the Hantush leaky equation to estimate the drawdown (vacuum) associated with operating vapor extraction wells at this flow rate.

The first step is to compute the gas conductivity for the sediment. Assuming, as before, that the hydraulic conductivity is 30 ft/day, the gas conductivity can be computed as follows:

$$K_g = K_w \frac{\rho_g}{\rho_w} \frac{\mu_w}{\mu_g}$$

where K_g = gas conductivity, in ft/day
$\quad K_w$ = hydraulic conductivity, in ft/day
$\quad \rho_g$ = gas density = 0.013
$\quad \rho_w$ = water density = 1
$\quad \mu_w$ = viscosity of water = 10,000 micropoise
$\quad \mu_g$ = viscosity of air = 183 micropoise (at 68°F)

In this example,

$$K_g = 30 \left[\frac{0.0013}{1} \bullet \frac{10,000}{183} \right]$$
$$= 2.13 \text{ ft/day}$$

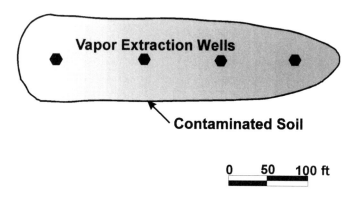

FIGURE 20.3. Location of four vapor extraction wells.

This makes the gas transmissivity $2.13 \times 40 = 85.2$ ft²/day.

In applying the Hantush leaky equation for computing vacuums, we will assume a leakance value for the surficial materials at the top of the vadose zone of 0.1 day⁻¹, a reasonable value. According to Hantush, drawdown around a leaky flow system can be expressed as follows:

$$s = \frac{3.58Q}{T} K_0\left(\frac{r}{B}\right)$$

where s   = vacuum, expressed in *ft of air*
        Q   = flow rate, in cfm
        T   = gas transmissivity, in ft²/day
        K_0 = Bessel function of the second kind and of zero order
        r   = distance from center of well, in ft
        B   = a leakage factor = $\sqrt{T/l}$
        l   = leakance, in day⁻¹

To calculate the vacuum just outside the borehole of a vapor extraction well, this equation is used with r set equal to the borehole radius, assumed to be 0.41 ft (10-inch diameter borehole). For this value of r, vacuum is computed as follows:

$$s = \frac{3.58 \bullet 65.5}{85.2} K_0\left(\frac{0.41}{29.2}\right)$$
$$= \frac{3.58 \bullet 65.5}{85.2} \bullet 4.382$$
$$= 12.06 \text{ inches of water vacuum}$$

Assuming the vapor extraction well efficiency will range between 20 and 80%, the calculated vacuum inside the well would be expected to range between 15.1 inches of water (80% efficiency) and 60.3 inches of water (20% efficiency). There would be an additional vacuum component caused by interference by the adjacent wells, but this contribution is minor and may be ignored.

Commercial blowers are readily available to sustain the desired yield at either of the calculated vacuum values, so the projected design is a good one, assuming reasonable well efficiencies are obtained. The Hantush equation can also be used to demonstrate that adequate vacuum is achieved everywhere within the contaminated area.

Thus, the target of 2 pore volume exchanges per day can be realized under the assumed conditions. Comparing 2 air exchanges per day with 1 water exchange per 98 days, it is clear we have a 196-fold increase in the rate of pore volume exchanges with the vapor extraction system.

We expect SVE and Air Sparging to clean a site in 6 to 18 months. We expect pump and treat to take between 5 and 20 years. As can be seen, pore volumes exchange are a significant part of the advantage. Tighter soils will give air a larger advantage, and these advantages will exist even in saturated zones; i.e., Air Sparging. Of course, chemical properties can add to the advantage. Most chlorinated compounds are more volatile than they are soluble. But pore volume exchange is still the major advantage of air.

Limitations

ONE OF THE MAIN WEAKNESSES of pump and treat systems is that the water travels along preferred paths in the aquifer. This can be the result of macro or micro lenses in the soil material. Contaminants that do not come in direct contact with the water must diffuse over to an area that has water movement before they can be removed from the aquifer. It is this diffusion process that creates the flattening of the concentration curve at the end of the project. Diffusion controls the final removal of the contaminants, and is an important part of determining whether the concentration of the contaminants finally reaches the cleanup goal. This same geological limitation occurs with air as the carrier. The geology doesn't change just because the zone is unsaturated. There are still areas of preferred flow and relatively stagnant areas in VES systems. Diffusion still controls the end of the project.

Air Sparging has also been shown to suffer some of the same limitations as pump and treat. Recent tests have proved that the air moves through channels in the aquifer. These channels do not change over the course of the project even if the system is turned off and then on again. The air has preferred paths that it travels. Any contaminants that do not come into direct contact with those channels have to diffuse to a channel in order to be removed from the aquifer by the air carrier. As with water movement, this effect creates the same flattening of the concentration curve at the end of the project. Diffusion still controls the final removal and helps to determine whether the contaminants reach the cleanup goal. The air is able to remove more mass over a short period of time, but in the end, diffusion still controls the final concentration in the aquifer.

Other In-Situ Techniques

How MANY OTHER IN-SITU techniques are misnamed? Almost every method that we use relies on air or water as carriers. The new methods that use steam or surfactants use air or water to deliver the heat or chemical compounds and to remove the contaminants. Other methods may deliver their enhancement by a separate means, but then rely on one of the carriers to actually remove the contaminants from the ground.

The only true in-situ processes (the title was just to get your attention) still rely on carriers. Biological reactions in the unsaturated zone and the aquifer are mainly limited by the lack of oxygen and nutrients. We use water and air to deliver these compounds to the contaminated area of the aquifer. Water or air carry the oxygen and nutrients to the bacteria so that they can use the contaminants as a food source and destroy them below ground. The mass transfer advantage of air and the diffusion limitations of both still have an effect on the design of an in-situ biological remediation.

When we look at remediation in this manner, we realize that there are not 20 or 30 new techniques; there are only two carriers with 20 or 30 methods to enhance their performance. The problem is that a large part of the performance of these carriers is limited by geology, not technology. Unless the new technique directly affects the geology, the performance will be limited by the carrier. The enhancement may improve the initial mass transfer, but will it overcome the diffusion limitations of reaching the final cleanup goal?

The best way to analyze any new remediation technique is to break the project up into two main phases. The first phase is the mass removal phase. How fast and at what cost does the remediation technique remove most of the contaminant from the ground or aquifer? Removing a high percentage of the mass in 6 months versus 10 years is an important economic and environmental advantage. This improvement alone may make it worthwhile to use the new technique. The second phase of the project is reaching the cleanup goal. The problem is that we usually only have time to analyze the mass removal portion of the project. We rarely determine if the aquifer will reach a 1 ppb level with the new technique, and how long it will take. The new technique may cost substantially more than standard methods, remove most of the mass in a short period of time, but still be limited by diffusion in the end. One solution may be a combination of techniques.

Since we will not have all of the performance data for several years (even from a pilot plant test), the only way to design an innovative treatment remediation is to understand the basic principles of the technologies. One principle that is consistent throughout all of the technologies is the use of carriers for delivery and removal. Mass transfer and diffusion limitations will be a part of every project. Understand and design for them.

21

PHYTOREMEDIATION

Evan K. Nyer and Edward G. Gatliff

INTRODUCTION

ONE OF MY COMPLAINTS ABOUT the environmental field has always been that new technologies come along for remediation of sites that have certain advantages over technologies that have been used in the past, and all of a sudden everyone decides that they have to use this technology or they are not "state of the art" in the remediation field. The cycle continues and the technology is overused and applied in improper locations and finally obtains a checkered reputation. Sparging and horizontal wells are probably the two "hot" technologies today. The problem is that both are entering their "overused" stage of application.

I'm going to predict that phytoremediation will be the next "hot" technology for the environmental remediation field. Phytoremediation is the use of plants to facilitate the destruction and removal of compounds from groundwater and soil. It is also referred to as vegetative remediation and one proprietary technique for the application of trees is called TreeMediation®.

Plant species can be selected to extract and assimilate or extract and chemically decompose target contaminants. Many inorganic chemicals considered environmental contaminants are, in fact, vital plant nutrients that can be absorbed through the root system for use in growth and development. Heavy metals can be taken up and bioaccumulated in plant tissues. Organic chemicals, notably pesticides, can be absorbed and metabolized by plants, including trees. The proper applications of these properties can lead to a very powerful remediation technique.

I have asked Edward G. Gatliff, Ph.D. to assist me in presenting this new area of remediation. Dr. Gatliff has years of experience in this area, and hopefully between the two of us we can provide everyone with the basic knowledge for applying this new technology.

APPLICATION OF PHYTOREMEDIATION

THERE HAVE BEEN FOUR MAJOR areas that have successfully used vegetation to help remediate contaminants. The following provides a quick review of each.

Uptake of Nutrients

THE OLDEST USE OF PHYTOREMEDIATION has been the application of plants to treat compounds that are considered nutrients for the plants. We all recognize that nitrates and ammonia in the groundwater can be considered a contaminant. Many municipal systems around the country have lost major sources of groundwater due to high nitrate concentrations. All plants require nutrients, and especially a nitrogen source in order for them to grow. If the root system of the plants can be put in contact with the groundwater contaminated with the nutrient, then the plant will remove the nutrient from the groundwater, resulting in a faster growing plant and clean groundwater. The example in the Field Data section will provide an example of this use.

Uptake of Nonessential Metals and Organics

IT HAS ALSO BEEN LONG recognized that plants will uptake nonessential metals and organics into their structures. For example, water hyacinths have long been used as part of a wastewater treatment scheme to remove metals from wastewater before discharging to rivers and lakes. These uptake properties exist in many plants and trees and can be used to remove metals and organics from groundwater and soil. Several types of metals have been removed by these techniques. Pesticides have been the main organic contaminant that has been successfully removed by plants from the environment.

Creating an Environment of Diverse Aquivial Population

THE NEXT AREA THAT HAS been found to be useful is the actual environment that the root system of the plants set up underground. It has been shown that the root system is an excellent location to grow a diverse group of microorganisms. The root system also creates an aerobic environment for biological destruction. As we have discussed in previous columns, I believe it is more important to set up the correct environment to be able to destroy organic compounds below ground than it is to simply name a specific microorganism that degrades a compound. Trees and other plants will be a very important part of creating the correct environment for the microbiological population to flourish and destroy the compounds that are of interest.

Water Pumping Action

THE UPTAKE OF WATER BY trees can substantially influence the local hydraulics of a shallow aquifer, thus controlling the migration of a contaminant plume. This "pumping" effect flushes water upward through the soil column and can be much more effective at remediation than traditional pump and treat systems.

It has long been recognized that phreatophytic (plants that are known for fast growth and high water usage rates) trees are effective at rooting very deeply—to 100 feet and more. In fact, phreatophytes have been studied as a nuisance in the semiarid to arid regions of the western United States where water is scarce. These plants are known to affect water availability by significantly lowering groundwater levels. Diurnal fluctuations of wells were reported in the 1940s and attributed to a grove of nearby cottonwood trees (USDA, 1955). Aquifer levels at other locations were reported to have dropped 5 feet during the growing season due to water consumption by phreatophytes. It has been reported that a single willow tree "uses and loses over 5,000 gallons of water in one summer day" (Miami Conservancy District, 1991). This seemingly phenomenal figure is comparable to what 0.6 acres of the phreatophyte alfalfa can transpire in one day (Schwab et al., 1957) and is plausible when the leaf surface area of a fairly large willow tree is considered.

At a site in southwestern Ohio, cottonwood trees demonstrated considerable pumping capacity, even in a relatively humid environment. A fairly ideal situation was available where two 40-foot-tall cottonwood trees could be isolated and evaluated. Monitoring wells were placed around the cottonwood trees and monitored for an entire season. Figure 21.1 clearly denotes the onset of transpiration, the drawdown of the aquifer during the season, and the rise in the elevation of the aquifer as the trees approached dormancy. Additionally, there was a downgradient trough, as demonstrated by the difference between the upgradient and downgradient well elevations. Calculations based on the rate of drawdown suggest the pumping rate for each cottonwood tree ranged between 50 and 350 gallons a day.

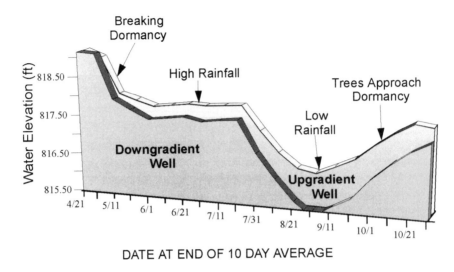

Figure 21.1. Aquifer drawdown by cottonwood trees.

FIELD DATA

While I refer to this technology as new, the original applications are over five years old and the understanding of the process and potential for use with organic compounds is over 50 years old. Some of the data from the field applications follow.

Ammonia and Nitrate

In 1990, phytoremediation was adopted as the methodology of choice to control a nitrogen-contaminated aquifer in New Jersey that could not be pumped efficiently. Soil conditions were such that the aquifer could be considered an aquiclude. However, groundwater was still migrating off-site and had to be contained. Various treatment alternatives were considered, including multiple pumping wells and a collection trench. These proposals were unsatisfactory due to the uncertainty of their effectiveness in this situation, as well as their cost. Interception and conventional treatment would have required significant long-term expenditures. Phytoremediation was ultimately chosen as a cost-effective method of treatment.

At this site, the soil conditions were very much a potential impediment for deep root penetration. In addition, the aquifer at this site was located at about 16 feet below the surface—a push for even a deep-rooted crop like alfalfa.

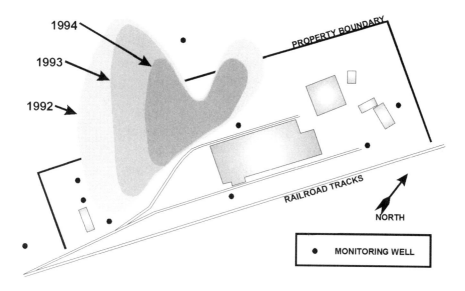

FIGURE 21.2. 50 ppm nitrate plume.

Specifically, the problem was to obtain rooting activity to a depth of 16 feet in a "tight" soil. Trees became the focus of attention.

Groundwater monitoring from 1992 through 1994 (four growing seasons) indicates that the off-site migration of the nitrogen plume (both ammonium and nitrate) has been affected and the apparent contraction of the plume is occurring (Figures 20.2 and 20.3). These data not only demonstrate the remedial value of the trees, but also the potential impact that trees can have on the hydraulic control of groundwater, especially in constrictive geology.

Nitrogen removal by the trees was determined to range between 40 and 80 pounds of nitrogen per acre for 1993. At full canopy, these values easily approach removal rates of 200-300 pounds per acre. Known removal rates of water and nitrogen by an acre of trees could be projected to consume an 8-16 foot thick nonreplenished aquifer (assuming 25-50% pore space) containing 30 mg/kg nitrogen in a single year.

Metals

AS PREVIOUSLY INDICATED, PLANTS HAVE been and are being used to clean up heavy metals in soil and groundwater. The potential for using hybrid poplar trees for heavy metal attenuation and extraction was enhanced when leaf samples of trees observed to be stunted in growth were determined to contain

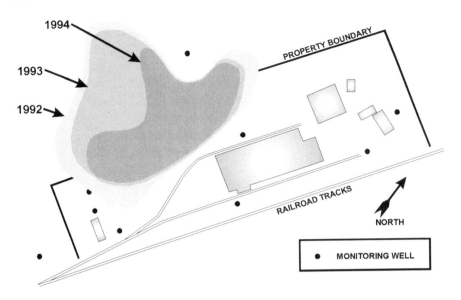

FIGURE 21.3. 50 ppm ammonium plume.

relatively high levels of zinc (up to 4,500 mg/kg-dry wt). This information led to a greenhouse experiment performed by Cristina Negri and Ray Hinchman, Ph.D., Argonne National Laboratory, and Dr. Gatliff, Applied Natural Sciences, Inc., through a Cooperative Research Agreement.

While the evaluation of all the data of the study is not yet complete, significant results are apparent. Briefly, pots were watered with solutions high in zinc throughout the growing season. No phytotoxic effects were observed even for the highest levels of zinc application. Mass balance calculations of zinc content in the pot leachate and leaf tissue did not adequately account for the amount of zinc applied to the system (a control pot root system of these trees have a relatively enormous capacity to attenuate zinc. Tissue analyses of root segments were determined to contain zinc concentrations as high as 50,000 mg/kg-dry wt. These apparent data lend significant support to the potential for using trees to attenuate heavy metal migration into the groundwater.

Organics

USING TREES AND OTHER VEGETATION of organic compounds is undoubtedly the fastest growing area of phytoremediation application. The plant's ability to absorb, assimilate, and transpire soluble or co-solubilized organic compounds has been demonstrated for over 50 years through herbicide and insecticide research. Studies in recent years by McFarland, Strand, and Anderson and Walton, to mention a few, illustrate the uptake, metabolism, volatilization, and collateral degradation of a wide range of organic compounds.

The state of research for this technology has evaluated the fate of only a few organic compounds by specific plants. Most notable for trees has been the work by Strand, who demonstrated that hybrid poplars can uptake, metabolize, and volatilize TCE. While more research is needed and preferred, the technology does not have to be stifled with regard to potential applications. A generalization for applying this phytoremediation technology would hold that the more soluble the organic compound, the greater the probability that the plant can extract it from the soil solution or groundwater. Less soluble, soil bound contaminants can also be treated through the development of a root system in the area of contamination. The rhizosphere (the area around the root system) has been shown to be a source of diverse microbial populations (naturally occurring or inoculated) that can increase the degradation of many less soluble compounds, including PCB.

Considering the relatively low cost of phytoremediation and the absence of viable alternatives, we will undoubtedly find that field data will occur in advance of research data for specific plants and contaminants.

ECONOMICS

Cost Comparison

AT A SITE IN ILLINOIS, TreeMediation was coupled with a pump and treat system to mitigate an immediate "at-risk" situation. In the initial phase of this remedial activity, TreeMediation will function primarily as the "treatment" phase of the pump and treat system. Eventually, TreeMediation will assume both phases of the system. Fortunately, this situation provides a unique opportunity to compare the costs of the two systems. The estimated costs provided below were developed based on this experience and are intended for illustration purposes only.

Pump and treat and TreeMediation costs were estimated in round numbers for a 1-acre site with an aquifer 20 feet deep. Costs common to both approaches were not included, such as meetings with regulators and laboratory analysis. Table 21.1 gives the items and costs considered for the pump and treat system and the TreeMediation system. It was assumed that the pump and treat system could function with three pumping wells and a reverse osmosis system for treatment. As Table 21.1 shows, the estimated costs for the TreeMediation program are considerably lower than those for the pump and treat system.

Table 21.1. Estimated Costs–1-Acre Site with 20-Foot-Deep Aquifer

Pump and Treat[a]

Equipment	$100,000
Consulting	25,000
Installation/construction	100,000
5-year costs	
Maintenance	105,000
Operation	50,000
Waste disposal	180,000
Waste disposal liability	100,000
Total	**$660,000**

TreeMediation

TreeMediation program design and implementation	$50,000
Monitoring equipment	
Hardware	10,000
Installation	10,000
Replacement	5,000
5-year monitoring	
Travel and meetings	50,000
Data collection	50,000
Annual reports	25,000
Effectiveness assessment–sample collection & analysis	50,000
Total	**$250,000**

[a] Assuming off-the-shelf equipment, three pumping wells, and a reverse osmosis treatment system.

Obviously, these costs can vary widely depending on specific circumstances. However, when the use of TreeMediation is a viable option, it should be a more cost-effective approach than pump and treat. There is another important consideration: additional disposal of pump and treat residues can result in additional regulatory problems. Usually, treatment with vegetation results in no disposable residue.

CONCLUSION

AS CAN BE SEEN BY all the above examples, phytoremediation is going to be an important part of controlling and remediating contaminated sites. This process is cost-effective, technically effective, and obviously environmentally favorable. The important thing to remember is that this is still only one technology to be used in our remediation designs. It cannot be applied in all situations for all compounds. It should also be remembered that this technology is not simply the buying of plants from the local K-Mart and placing them in the soil near a contaminated site. There are very specific design requirements for the proper application of this technology and I suggest, as always, that someone with specific experience in this area be part of the team that provides the remediation design. Hopefully, we will all use this new technology to its full capabilities without making it the next "hot" environmental technology and overapplying it. I would like to see this technology maintain the reputation it is going to obtain over the next several years. We probably will revisit this technology in the future as new applications are found.

The nostalgic part of this technology is that I entered this field 20 years ago as an environmentalist. Back then we were called tree huggers. After using all the plastic, steel, and mechanical methods to remediate sites, it is nice to be able to go back to hugging trees.

REFERENCES

California Fertilizer Association (Soil Improvement Committee). *Western Fertilizer Handbook,* 5th ed. Interstate Printers and Publishers, 1975.

Miami Conservancy District, *Aquifer Update,* No. 1:1, 1991.

Schwab, G.O. et al. *Elementary Soil and Water Engineering,* 2nd ed. John Wiley & Sons, New York, 1957.

U.S. Department of Agriculture. *"Water," The Yearbook of Agriculture 1955.* Washington, DC, U.S. GPO, 1955.

22

PLUMES DON'T MOVE

Evan K. Nyer and Mary J. Gearhart

ONE OF THE PROBLEMS THAT we all have when we design remediation systems is that we rely on one underlying assumption: the contamination plume is moving. The purpose of this article is to propose that this assumption may not be true. And, more importantly, that the better way to design remediation systems is based upon the concept that the plume is stable. I can see that skeptical look on all of your faces. However, before you stop reading, remember who is writing this article. I have presented some very new (strange) ideas before in this column, and have been able to convince some of you to use them. However, knowing that this is going to be a tough sell, I have asked Mary Gearhart to help me with the presentation. So give us a page or two before you decide on the value of this idea.

The first thing to consider is that there is only one main process that makes the plume move—advection. There are six or more processes that limit, prevent, or stop the movement of plumes. Any time that the movement due to advection is equal to the negative effects from the other processes, then the plume stops moving.

Before we start to discuss these processes, a couple of ground rules need to be established. First, when we talk about the plume stopping, we are referring to the plume as represented by a low concentration limit (e.g., <5 ppb). If we could measure it, there is always the possibility that a single molecule of the contaminant keeps moving.

The second rule is that, like all good rules, we are not trying to describe every situation. There are always geologic conditions such as fractures, karst geology, etc., plus man-made conditions such as surface structures, and buried utility lines that will affect the movement of the water and the interaction between the contaminants and the geologic soils through which the plume moves.

Let us look at each major process that occurs with the plume movement, and in the end summarize the processes that make materials of the plume move vs. the processes that limit the movement of the plume constituents.

The Plume Moves–Advection, Dispersion

THE MAIN MECHANISM THAT ALLOWS inorganic and organic contaminants to move in an aquifer is the movement of the carrier, water. As we have discussed many times, the water picks up the compound and moves it in the direction controlled by gravity and geologic conditions. The amount of contaminant that the water can carry is controlled by the solubility of that particular compound.

Solubility is one of many properties of a chemical. Solubility is expressed as the number of milligrams of a constituent that can dissolve in one liter of water (mg/L). For inorganics, solubility depends on the form of the constituent in solution, other dissolved constituents, pH, and oxidation-reduction potential. For organic substances, the solubility is the mass of that organic substance that will dissolve in a unit volume of water under specific conditions. For purposes of comparison, the solubility in water of quartz is 12 mg/liter while that of trichloroethene (TCE) is 1100–1500 mg/L (Montgomery and Welkom, 1990). That means that under the same temperature, pressure, and pH, more TCE will dissolve into water than quartz, or in other words, the water will carry more TCE than quartz. In a groundwater flow system, a chemical with low solubility will generally not migrate a significant distance from its source, while a chemical with the high solubility is more likely to do so.

Advection is controlled by the carrier (usually water) and geology. That is, the dissolved particles are physically indistinguishable from the water particles and move as if they are water. Advection is the primary component of water and particle movement, and accounts for the longitudinal flow of the plume.

Dispersion is an important, yet smaller component of contaminant or plume movement. Dispersion is the result of mechanical mixing (velocity differences within the pathway) and molecular diffusion (differences) in kinetic activity between ions. Both of these processes are grouped broadly into a category called hydrodynamic dispersion. Together, advection and hydrodynamic dispersion represent processes that appear to "move" the plume forward.

The Plume Doesn't Move—Geology, Hydrogeology

THERE ARE SEVERAL PHYSICAL CONDITIONS that limit the movement of the carrier itself or the compounds within the carrier. Natural geologic conditions (e.g., high density soil materials like clay) can prevent the water from moving through specific zones. If the source area is within one of these geological zones, then the groundwater will move around the tight soil and not directly interact with the contaminants. Without a carrier, the movement of the contaminants will be very small. Diffusion will be the main process that moves the contaminants within the tight soil. Advection will take over once the compounds reach the limits of the tight soil and enter the main flow paths of the groundwater. The key factors relevant to this discussion are porosity and hydraulic conductivity.

In addition, new water is constantly added to the aquifer. This new water dilutes the concentration of the contaminants. This water can be from the surface in the form of rain or other mechanisms that allow water to travel down through the vadose zone or up from deeper zones into the aquifer. Water must move from its source through the medium. Impediments to that movement start at the top of the hydrologic cycle. Many texts have been written to describe these features and examine them in great detail. For our purposes, it is important to understand that those processes serve to slow the water and save us from the "underground river" theories. Most contamination is a point source and as the plume spreads in the direction of movement, the width of the plume increases bringing natural dilution.

The Plume Doesn't Move—Geochemistry

THERE ARE THREE MAIN MECHANISMS that can stop a compound within the plume from moving based upon geochemistry: adsorption, precipitation, and ion exchange (a subcategory of sorption mechanics).

Adsorption is one of the simpler processes to describe relative to plume movement (or lack thereof). Adsorption occurs when a molecule of the contaminant in the plume is attracted or attached to a soil particle within the aquifer. Adsorption capacity is controlled by the nature of the soils as well as the contaminant properties. Solubility, octanol-water partition coefficient, and the molecular structure of the contaminant are all important factors in this process (Nyer, 1992). The amount of soil surface area, the ionic charge, and its organic content determine where and if adsorption can occur (adsorption capacity) and the contaminant properties determine how much material is available to adsorb. Adsorption can play a significant role in delaying the movement of the plume.

The second geochemical characteristic that affects plume movement is precipitation. Precipitation applies only to inorganic materials and is a chemical
reaction. The inorganic material is in solution (dissolved) and the adjustment
of the ion concentrations (e.g., a change in pH) causes the material to dissoci-
ate or form a "solid" or precipitate. Once the contaminant is in an insoluble
form it is no longer part of the groundwater. With proper sampling techniques, the precipitate should not be part of any groundwater sample. In most
cases, this means that the groundwater is no longer contaminated and the
plume ceases to exist.

The third process, ion exchange, is closely related to adsorption. The soils
have an outer layer of ions that react to the water and contaminants "moving"
by them. The ionic charge on this outer layer changes based on the ionic
strength of the passing materials. When the conditions change enough, the
ions on the outside layer can be exchanged or substituted. Generally, ion exchange occurs in soil matrices that are largely clay. So, again, this process
affects passing contaminants which have the appropriate molecular makeup.

The Plume Doesn't Move—Biogeochemical Reactions

MORE IMPORTANT THAN SIMPLE LIMITATIONS to the plume movement are several processes that actually can destroy the compound in the aquifer environment. While the most famous of these processes are biological in nature, we
must also include the purely chemical reactions (i.e., abiotic) that occur below ground that can also destroy compounds.

The main biological reactions that occur below ground can be fitted into
two categories: aerobic and anaerobic degradation. Almost all organic compounds go through some degree of degradation processes while in the ground.
The rate of degradation relates to the type of compound, the number of substitutions or the degree of complication of the compound, and the environment in which the plume is moving. The general family of compounds of
alcohol, ketones, and petroleum hydrocarbons are mostly aerobically degradable. Compounds such as chlorinated hydrocarbons have been found to be
affected by a bacteria under anaerobic or reducing conditions. Complex compounds such as the creosote family or PCBs have differing rates of degradation depending on their level of complexity.

Petroleum hydrocarbons are probably the best known compounds for natural degradation. It is becoming common practice across the country to include
biodegradation as part of the remediation concepts employed for gas station
and terminal releases. The best two recent summaries of the concept are provided by the AFCEE and Lawrence Livermore reports (Wiedemeier et al.,
1994; Rice et al., 1995).

There is also a significant amount of work on the degradation of chlorinated compounds. This family of compounds is most concerned with perchloroethylene, trichloroethylene, etc. Researchers have found two main mechanisms for the disappearance of these compounds in the aquifer environment, dehalogenation and co-metabolism. The dehalogenation occurs mainly under highly reducing conditions. The bacteria produce enzymes that remove the chlorine or other halogen compound off the base carbon. This leaves a carbon compound stripped of its chlorine molecules. It can then undergo natural degradation processes as discussed under the petroleum hydrocarbon section. The other main mechanism is co-metabolism. In this process, a degradable compound is used by the bacteria to grow and thrive. However, under the right environmental conditions with the right bacteria and correct co-metabolite, these bacteria also produce enzymes that attack the chlorinated compounds. The main processes in this area that have been studied have been methane and ring compound addition to stimulate the bacteria that produce the necessary degradation enzymes. More details on this area are also available in the literature (Nyer, 1992; and Nyer et al., 1996). In addition, the EPA has just published a protocol for studying degradations of chlorinated hydrocarbons (Wiedemeier et al., 1996).

The Plume Doesn't Move—Chemical Reactions

THE OTHER PROCESSES THAT MUST be remembered when determining the natural length of an organic plume are the chemical processes. Although people concentrate on the biological processes, it must be remembered that chemical processes also occur below ground. Probably the most recognized chemical process is hydrolysis. I have worked on one project in which that was the controlling destruction mechanism. Hydrolysis was the main process that gave a biologically nondegradable compound a half-life of six years in the aquifer. These processes can be very important on certain compounds and must be included in the total understanding of the contaminant movement below ground.

The Plume Doesn't Move—Equation

EACH OF THESE FACTORS CAN then be combined into an equation. The equation is made up of the positive factors that make the compounds move and the negative factors that restrict the movement or decrease the presence or concentration of the compounds themselves. When the positive factors equal the negative factors, the plume stops moving. At some distance from the original source of release all of the factors come into equilibrium; therefore, PLUMES DON'T MOVE!

Plume Movement = Advection + Dispersion – Dilution – Retardation – Precipitation –   Ion Exchange – Biochemical Degradation – Abiotic Degradation.

The real question is "Has the plume reached equilibrium?" and if it has not, then "Why not?" It is the "why" that really controls our remediation designs.

The Plume Doesn't Move–Who Cares?

THE PURPOSE OF A REMEDIATION design is to control and clean a contaminated aquifer. We all now realize that we will not be able to return a contaminated aquifer to pristine conditions. Our design goal is to get the aquifer to a level that no longer threatens human health, or the environment as defined by cleanup levels and/or risk assessments. The main aspect of this process is to reduce the concentrations in the aquifer and to limit the pathways in which the remaining compounds may contact humans. How does the above discussion relate to this process?

One of the first questions that we always ask during a remedial investigation is if the plume is still moving. Most regulators require that the plume be controlled. No matter what we are doing to clean the source, the plume must not move to a point where it can affect people. This also becomes very important when the plume is approaching the property line or other institutional boundaries.

Normally the first thought is to use advection to stop the movement of the plume. We can do anything from putting a wall around the contaminant to pumping the groundwater in order to stop the natural flow pattern. Any pumping within the contaminated zone will require aboveground treatment.

The purpose of this discussion is to show that there are other processes that can stop the movement of the plume. We should also be thinking about any of the negative factors in the equation when trying to design.

The other important question is, "Do you really understand the plume unless you understand all of the negative effects on contaminant movement?" Too often we settle for information on advection and retardation as our only information collected during the remedial investigation. There are too many other important processes active in the aquifer for this limited information to provide an accurate picture of the site. We will not understand how to design unless we really know what the problem is. The designer should know all of the negative effects, not just retardation, in order to design a cost-effective remediation. If you start with the premise that plumes don't move, you will have a better chance of incorporating all of these significant design factors into your remediation.

The controlling factor becomes the "why" in the plume moving. If there is a sand lens or fracture, then advection may be the best process. If the pH is very low at the source, which prevents precipitation and ion exchange, then a correction of the pH may be the preferred remedy. We can also change the environment in the aquifer to enhance a natural negative effect. Examples that we have discussed in previous columns would be reactive zones and enhancement of biological activity. Either of these processes could be used to remove compounds as they move in the aquifer. When the negative effects are enhanced, then the plume will be shorter or controlled at its present location without the use of pumping. This is good design.

REFERENCES

1. Montgomery, J.H. and L.M. Welkom. *Ground Water Chemical Desk Reference.* Lewis Publisher, Boca Raton, FL, 1990.
2. Nyer, E.K. *Groundwater Treatment Technology,* Second edition, Van Nostrand Reinhold Publishers, 1992.
3. Wiedemeier, T.H., M.A. Swanson, S.W. Moutoux, J.T. Wilson, D.H. Kampbell, J.E. Hansen, and P. Haas. "Overview of the Technical Protocol for Natural Attenuation of Chlorinated Aliphatic Hydrocarbons in Ground Water Under Development for the U.S. Air Force Center for Environmental Excellence." AFCEE. EPA 540/R-961509, 1996.
4. Rice, D.W., B.P. Dooher, S.J. Cullen, L.G. Everett, W.E. Kastenberg, R.D. Grose, and M.A. Marino. "Recommendations To Improve the Cleanup Process for California's Leaking Underground Fuel Tanks," Lawrence Livermore National Laboratory, University of California, October 16, 1995.
5. Nyer, E.K., D. Kidd, P. Palmer, T. Crossman, S. Fam, F. Johns, II, G. Boettcher, and S. Suthersan. *In-Situ Treatment Technology,* CRC Press/Lewis Publishers, Boca Raton, FL, 1996.
6. Wiedemeier, T.H., D.C. Downey, J.T. Wilson, D.H. Kampbell, R.S. Kerr, R.N. Miller, and J.E. Hansen. "Technical Protocol for Implementing the Intrinsic Remediation with Long-Term Monitoring Option for Natural Attenuation of Dissolved-Phase Fuel Contamination in Ground Water" Air Force Center for Environmental Excellence, Brooks Air Force Base, San Antonio, TX, August 29, 1994.

23

MAY THE FORCE BE WITH YOU

Evan K. Nyer, Gary Boettcher, and Pete Jalajas

DURING 1996 I USED ONE remediation technology more often than any other. It was not a very advanced technology, relying on simple movement of air and water across the contamination zone to remove the contaminants. There are not many articles written about this technology, there are no magazine covers devoted to it, and there have been no conferences focusing on the use of this method. However, I found it to be one of the most flexible and powerful technologies for the cost-effective and efficient removal of contaminants from the ground. I refer to this technology as the Vacuum Enhanced Recovery (VER). It is the combination of liquid and gas movement at the same time, in the same well.

Most readers will have heard of this technology referred to as "bioslurping." I hate the name bioslurping. Nonetheless, bioslurping is the use of a vacuum system lowered into the ground to remove LNAPLs, liquids, gases, and anything else that can be sucked out with the vacuum equipment. The bio part comes with the addition of air to the treated zone to increase biological activity.

My prejudice may be due to the fact that I didn't invent the name, but I think I also have some good technical reasons. Vacuum-Enhanced Recovery refers to combining the forces of gravity with the forces of pressure (vacuum) to drive the air and liquid to a point where it can be removed from the ground. The trouble with bioslurping is that this leads the designer to think of this technology in a very limited applicability. Simply thinking about what you can suck out of the ground with a vacuum tube severely limits the breadth of accomplishments that can be obtained by the combination of vacuum and pumping.

I know that I can't get everyone to stop using the term "bioslurping", however, the purpose of this article is to expand your thinking so that the technology, VER, can be applied to a broader range of projects. I have asked Gary Boettcher and Pete Jalajas to assist me in the writing of this article. They were the respective project managers on two of the VER designs and installations that were performed during 1996. The first one is a classic single pump vacuum removal that most people would refer to as bioslurping, with the second project showing the true breadth of this technology, using a dual pump system to remove material over 60 feet below ground surface.

VER DESIGN

ALTHOUGH VACUUM-ENHANCED RECOVERY HAS been used for decades as a standard approach for dewatering and stabilization of low permeability sediments or to speed dewatering of more permeable sediments, it has been only recently that it has been incorporated into groundwater remediation applications. The use of vacuum-enhanced recovery systems in environmental remediations is unique because whereas most remediation methods rely on either water or air as the carrier, vacuum-enhanced recovery uses a combination of two forces, gravity and pressure differential, to move the water. This can be very beneficial in enhancing cleanups when used in the proper hydrogeologic setting.

Before, we have discussed how gravity is used as a main force to move water in all methods that use water as a carrier. The main method to control the direction of water movement is to use a well to remove water from the aquifer. This creates a drawdown of the water in the aquifer; water travels from a place of high head (high water level) to a place of low head (low water level at the bottom of the pumping well). The liquid state of water allows this vertical force to create a horizontal movement of the water. This principle has been successfully practiced as a remedial technique in the medium and high permeability geologic formations. The success of this technique (using gravity alone as the main force), however, can be severely restricted in the low permeability formations due to the diminished groundwater flows that can be achieved by standard remedial recovery equipment. Other areas of limitations are perched aquifers and thin areas of a highly transmissive aquifer. Vacuum-enhanced recovery systems overcome this limitation by using a second force, pressure differential, to help the movement of the water when gravity movement is limited by the geology. Vacuum is applied, in addition to pumping, to move the water. This combination allows us to move air and water in geologic formations that were inaccessible before.

There are basically two types of systems: a single-pump system, which uses a combined liquid/vapor pump, and a dual-pump system, which uses separate liquid and vapor pumps. The single-pump system uses a single-pump with a drop pipe to withdraw both fluids and vapors. Because we are dealing with a suction lift system, we are limited to a maximum theoretical lift of about 34 ft. In practice, this type of system usually is limited to lifts of 15 ft or so and thus is applicable for shallow aquifers only. One of the side benefits of this design configuration is that the tube can be lowered into a LNAPL. This can result in LNAPL removal without the need for water drawdown and subsequent treatment. The benefits of a single-pump system are lower capital and lower operational and maintenance costs. In addition, since all the equipment is aboveground, it is easier to install and maintain. Disadvantages of the suction lift system include limited depth, the problem of balancing vacuum at multiple wells, and the higher vacuum required to maintain the lift to produce fluids, which limits design flexibility.

Two-pump systems can be used for greater depths. For these systems, a vacuum is applied at land surface and a second downhole pump is used to withdraw fluids. These systems are easier to balance and operate when multiple extraction wells (greater than five) are involved, and allow more design flexibility after selecting the optimum vacuum pressure. Disadvantages include the increased capital, operation, and maintenance costs of two pumping systems, and the care that must be taken in selecting a downhole pump that will operate under vacuum.

SINGLE PUMP VER (BIOSLURPING)

Tibbetts Road Superfund Site

Background

THE SITE IS A FORMER residential property covering approximately two acres in southern New Hampshire. It is in a forested rural residential neighborhood. The site is underlain by 20 to 30 feet of relatively permeable glacial till (upper overburden), below which is a similar thickness of very dense lodgment till (lower overburden). The depth to the uppermost waterbearing zone is approximately 5 to 10 feet below land surface; shallower in the spring, deeper in the fall. The hydraulic conductivity of the upper till is on the order of 10^{-3} to 10^{-4} cm/sec. The hydraulic conductivity of the lower till is on the order of 10^{-5} to 10^{-6} cm/sec.

The site was discovered in the mid-1980s with 400 drums of mismanaged chemicals. The groundwater contains actionable levels of chlorinated solvents

(TCE, 1,2-DCE, PCE), aromatics (BTEX), and ketones (MIBK), among others. These compounds were individually present up to single- and double-digit parts per million concentrations. Three small plumes of compounds emanate radially from the central area of the property. The USEPA performed source removal of the shallow vadose zone contaminated soils, but they did not address saturated soil and groundwater.

Strategy

OUR OVERALL REMEDIAL ENGINEERING STRATEGY involved the following fundamental principles. Groundwater was to be removed from the subsurface only to the extent that it removed water from the contaminated shallow saturated soil. Air was to be used as the carrier medium to maximize volatile contaminant removal and to minimize extracted groundwater during the initial, short-term, active, mass-removal, phase of the project life-cycle. The site was to be capped to minimize infiltration of rainwater and to expand the influence of the air removal system. The active phase extraction was only to be performed while such removal was most cost-effective; that is, until the mass removal vs. time curves became asymptotic. At that point, natural attenuation methods would clean up the residual vadose zone and groundwater, reducing the remaining groundwater concentrations to below the cleanup levels during the "reaching clean," asymptotic phase of the life-cycle.

Project Design and Installation

IN SPRING 1995, AN ASPHALT cap was installed over the entire two acres of the site. The VER system was installed soon thereafter. The VER system includes two skid-mounted, 5-horsepower, liquid-ring, single-phase electric pumps (manufactured by Atlantic Fluidics). The liquid-ring pumps remove both air and water from the subsurface. The extracted air and water from the wells are manifolded near the $10' \times 10'$ treatment plant area. After the knock-out tank, the vapor and liquid phases are then passed through separate granular activated carbon units. The VER pump and treatment system was installed in a central location of the site so that it could affect all three treatment cells with the minimum of hardware.

Due to the small size of the contaminated area, we were able to use the technique of expanding the pilot plant directly to the full-scale system. This saved us about one year on the project schedule and over $200,000 in reports and construction costs.

Based on prior data, one upper overburden extraction well was installed in May 1995 at a location inside the fenced portion of the site, forming Treatment Cell 1 for the overburden aquifer pilot test. In addition, five piezometers and five vacuum probes were installed near the extraction well to monitor the impact of Treatment Cell 1 on groundwater and the vadose zone. In August 1995, two additional extraction wells were installed at locations northeast of the existing extraction well to expand the area of influence of Treatment Cell 1. Three vacuum probes were also installed near these new wells to monitor the effectiveness of the expanded treatment cell.

In August 1995, two new treatment cells were created at the site (see Figure 23.1). The locations of these cells were selected based on the results of a microwell investigation we conducted at the site, which refined the delineation of the three areas of impacted groundwater.

In July and September 1996, two additional extraction wells were installed within Treatment Cell 1 to expedite the remediation of groundwater in the Treatment Cell 1. These wells were installed because the mass removal from one of the extraction wells was not declining as rapidly as was desired. Two vacuum probes were installed near these new wells to monitor the effectiveness of the expanded treatment cell.

Operations

VER WAS INITIATED (AS A pilot test) in spring 1995. Depending upon the maintenance being performed, up to seven extraction wells were operated at any given time. Groundwater has been extracted from the individual extraction wells at an initial rate of about 2 gpm, declining to a sustained rate of about 0.5 gpm, under vacuum-enhanced conditions. Typical water level depressions produced by the system are shown in Figure 23.2.

Vacuum applied at the wellheads was typically 10 to 15 in. Hg. Air flow was zero at spring startup when the well screens were filled with water, and then increased to about 50 scfm total (up to about 10 scfm from each well) as groundwater was removed and air was able to enter the wells. Air concentrations of the key compounds in the most contaminated single well initially totaled 30 ppm at startup, increasing to about 80 ppm, declining again to about 20 ppm at the asymptote.

Figure 23.3 shows the removal of approximately 60 pounds of the regulated compounds in 1995. The mass removal rate peaked at 0.5 pound per day, dropping to about 0.1 pound per day late in the year. After a winter hiatus, VER was restarted in April 1996. During that treatment year, we removed another 10 pounds of regulated compounds for a total of about 70 lb

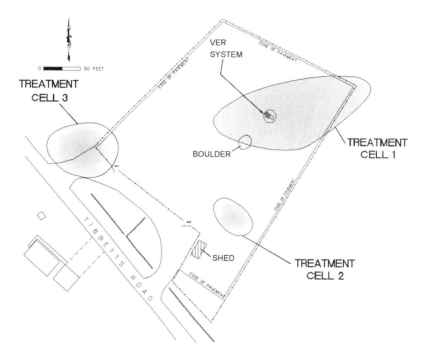

FIGURE 23.1. Treatment cell layout plan.

removed. The mass removal rate in 1996 began at 0.08 pound per day, dropping to and asymptoting at about 0.04 pound per day late in the year.

The system will be run for one more year with minor changes to the extraction wells. Natural attenuation will complete the remediations after the VER equipment is removed.

DUAL PUMP VER

Southern California Site

Background

DURING 1995 AND 1996, AN approximately 100 acre aerospace research and development, and manufacturing facility was demolished in preparation for residential redevelopment. The property is located in southern California and the Regional Board was responsible for authorizing cleanup requirements and approving closure.

Numerous soil and groundwater investigations had been completed prior to 1992. The results of the investigations identified an area of the property

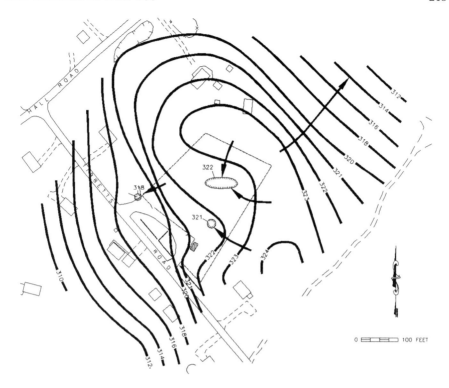

FIGURE 23.2. Overburden water levels, December 1996.

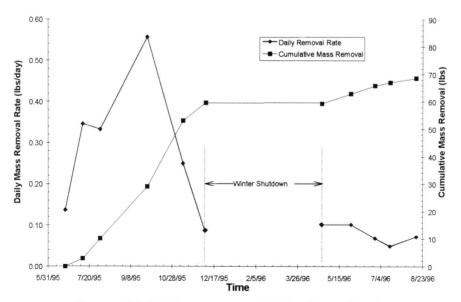

FIGURE 23.3. VOC mass removal: All cells combined.

where the groundwater was impacted with chlorinated volatile organic compounds. The primary constituent was trichloroethene.

The portion of the property with impacted groundwater is approximately eight acres. This area is underlain by approximately 50 feet of silty sands and sands with silts, which in turn are underlain by bedrock deposits which are predominantly semiconsolidated silts and clays. The primary source of groundwater is cultural, including leakage from subsurface utilities and irrigation. Approximately 5 to 15 feet of groundwater accumulates on top of the interface between the sands and the underlying bedrock.

Strategy

A STRATEGY WAS DEVELOPED WHICH expedited groundwater remediation and allowed the property to be redeveloped in 1997. The strategy was based on selecting the Best Available Technology (BAT) to remediate groundwater where the total VOC concentration exceeded 500 µg/L. The objective was to remove approximately 6 saturated pore volumes of groundwater and approximately 200 vadose zone pore volumes during 6 months. Using groundwater and air as carriers, the goal was to remove the majority of VOCs, thereby preventing further migration.

The BAT for this property was VER for groundwater dewatering and soil vapor recovery. This technology was selected because dewatering could be rapidly achieved. VER could then rapidly remove VOCs in what would then be an unsaturated zone.

Project Design

GROUNDWATER EXTRACTION AND VER WAS performed using 38 dual purpose recovery wells. The optimal spacing between wells was established to be approximately 80 feet. Individual groundwater delivery lines from extraction wells were connected to a common groundwater extraction header that delivered extracted water to the groundwater treatment system (granular-activated carbon). The treated water was then discharged to a National Pollutant Discharge Elimination System (NPDES) permitted outfall. Individual VER delivery lines were connected to a common header that delivered extracted vapors to the treatment system.

Operations

PRIOR TO PUMPING GROUNDWATER, THE thickness of the saturated zone ranged from 8 to 13 feet. At the beginning of the program, the groundwater system

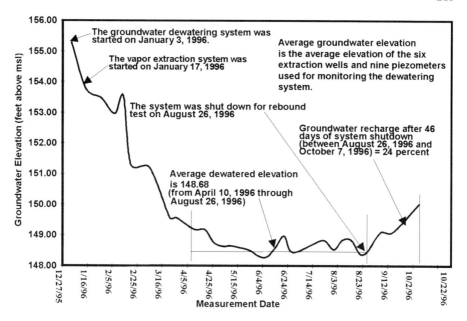

FIGURE 23.4. Average groundwater elevation vs. time.

extracted and treated approximately 25 gpm with the 38 pumps alone. The system was enhanced almost three-fold to 70 gpm when the wells were sealed and 60 inches (water) of vacuum was applied to the system. At the end of the program, the groundwater thickness average was three feet (Figure 23.4) and the flow rate declined to approximately 14 gpm. Based on these measurements, the groundwater zone within the well field was approximately 70% dewatered, and the water pumping was able to exchange approximately 5.7 saturated pore volumes.

Between startup of the remediation system and when the system was shut down, the combined groundwater VOC concentrations declined from 685 mg/L to 248 mg/L (Figure 23.5). The rate of daily VOC extraction from the saturated zone was approximately 0.5 pound per day. At the end of the program, the rate was 0.04 pounds per day. The cumulative amount of total VOCs removed via groundwater extraction was 32 pounds (Figure 23.6) and confirmation data demonstrated that the VOC concentrations did not increase once the system was shut down; in fact, the concentrations decreased.

The soil vapor combined flow rate ranged between 430 and 742 CFM. The vacuum at the wells ranged between 45 and 65 inches of water, which declined as the saturated zone was dewatered. The number of vadose zone pore volumes that were extracted was approximately 367, which exceeded the 200 pore volume target.

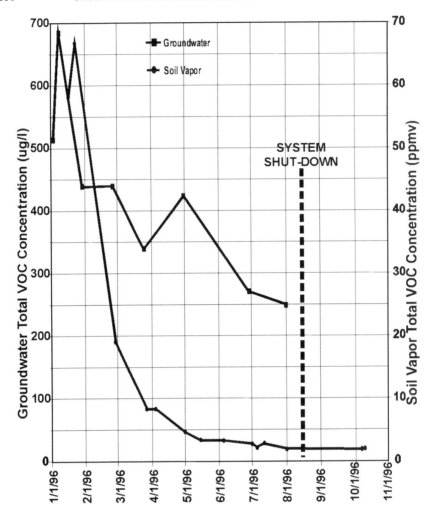

FIGURE 23.5. Total VOC concentrations.

Between startup of the VER system and shutdown, the combined VOC concentrations declined from 67 ppm based on volume (ppmv) to 2 ppmv, Figure 23.5. The VOC concentrations reached asymptotic conditions and did not increase once the system was shut down. The VOC extraction rate ranged from 14.7 pounds per day to 0.8 pounds per day, and the cumulative recovery was approximately 622 pounds (Figure 23.6).

The remediation program used BAT to remove approximately 6 saturated pore volumes and 367 air-filled pore volumes, and the VOCs were reduced to asymptotic concentrations in 8 months. Because the technology-based goal

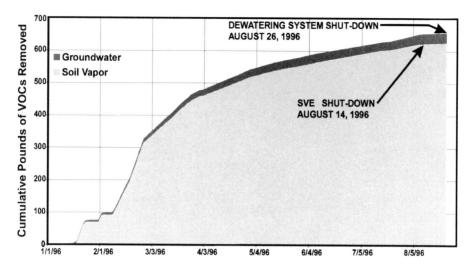

FIGURE 23.6. Cumulative pounds of VOCs.

was achieved and future groundwater source was eliminated, the Regional Board granted closure and property redevelopment was initiated.

SUMMARY

AS THE READER CAN SEE by these case histories, VER is a very powerful and flexible technology. By combining the second force of vacuum with the primary force of gravity we can move both air and water through our contaminated zones under a large variety of geologic settings.

If the reader would like more information and design methods for this technology they can refer to my book, *In Situ Treatment Technology,* which has a complete chapter on VER. May the VER Forces be with you.

LIST OF ACRONYMS

AS/VES	air sparging/vapor extraction system
BAT	best available technology
BTEX	benzene, toluene, ethylbenzene, and xylene
CoC	contaminants of concern
DCE	dichloroethene
DNAPL	dense nonaqueous phase liquid
DO	dissolved oxygen
EPA	Environmental Protection Agency
ESA	environmental site assessment
GAC	granular activated carbon
LTM	long term monitoring
MCLs	maximum contaminant levels
MEK	methyl ethyl ketone
NAPL	nonaqueous phase liquid
NGWA/AGSE	National Ground Water Association/Association of Groundwater Scientists and Engineers
NPDES	National Pollutant Discharge Elimination System
NRC	National Research Council
O&M	operation and maintenance
PCB	polychlorinated biphenyl
PCE	perchloroethylene
POTW	publicly owned treatment works
PRP	potentially responsible party
QA/QC	quality assurance/quality control
RFQ	request for quotation
ROD	records of decision
SVE	soil vapor extraction
SVOC	semivolatile organic compound
TCE	trichloroethylene
TDS	total dissolved solids
TPH	total petroleum hydrocarbon
TVO	total volatile organic
UST	underground storage tank
UV	ultraviolet

VC	vinyl chloride
VER	vacuum-enhanced recovery
VES	vapor extraction system
VOC	volatile organic compound

INDEX